Agile Software Development

T0254121

Thomas Stober • Uwe Hansmann

Agile Software Development

Best Practices for Large Software Development Projects

 Springer

Dr. Thomas Stober
IBM Deutschland Research &
Development GmbH
Schönaicher Str. 220
71032 Böblingen
Germany
tstober@gmx.de

Uwe Hansmann
IBM Deutschland Research &
Development GmbH
Schönaicher Str. 220
71032 Böblingen
Germany
uwe@hansmanns.net

ISBN 978-3-642-42557-8   ISBN 978-3-540-70832-2 (eBook)
DOI 10.1007/978-3-540-70832-2
Springer Heidelberg Dordrecht London New York

Cover design: WMXDesign GmbH, Heidelberg, Germany

Printed on acid-free paper

Springer is part of Springer Science+Business Media (www.springer.com)

"Life is what happens, when you are busy making other plans"
                                                        (John Lennon)

Foreword

Live with Heart and Soul- or Plan your Life?

Should you develop software in an agile manner or should you follow a detailed plan? Is your affection for the final result the primary motivation? Or is it rather the accurate processing of given steps listed in a project plan? I can assure you that dealing with these questions is almost like having two ideologies confronting each other. This makes it so difficult to establish agile software development within an organization as a principle, despite its remarkable successes.

Most of my time, I am dealing with innovation of all kinds. The confrontation between structure and flexibility exists in this context as well. Big enterprises want to manage their innovation in a precisely structured way. Innovation should emerge in a predictable way and based on an omniscient master plan. However, enterprises are surprised to realize that innovation happens spontaneously. It emerges in most cases directly from the lifeblood that the innovators put into their endeavor. Google and Amazon are big and profitable companies, while still preserving the agile spirit of their early days. Nevertheless, other enterprises only rarely attempt to copy this mentality. There is no plan for this.

It is quite depressing to know that the agile principle works better than traditional planning techniques, but often it does not receive the deserved respect or recognition. When I used to complain about this at home my father joked: "To be right, after all, does not mean that a right is acknowledged." Today, I am convinced that the dispute about the agile principle is deeply rooted in mankind and in the established understanding of management.

How can I explain this? Probably with the two human patterns "theory X" and "theory Y," which MIT Professor Douglas McGregor presented in 1960 in his book "The Human Side of Enterprise." Theory X assumes that mankind is lazy and without ambition to a large extent. Employees need to be offered an incentive for doing their work, and they need to be punished for their mistakes. They need precise instructions as to what they are expected to do step-by-step. Ideal would be a clear specification for each of their grips, similar to the work at a conveyer belt. Control and steering is done by managers. Managers are rather rare species, which are able to take over responsibility.

Theory Y assumes that actively pursuing challenging goals is a valuable aspect in everyone's life. To achieve a demanding goal, any effort will be undertaken. Humans will gladly take over responsibility, carry their point, and show self-discipline. Control and penalties are rarely necessary. Any occurring issues will be solved by ingenuity, persistence, and judgment. The responsibility of management is to remove obstacles and define a meaningful scope of work.

Theory X is closely related to the idea of scientific management, which emphasizes that there is one best and optimized solution to any problem. Theory Y is derived from humanistic management, which does not steer and control employees, but rather aims at coaching and developing people and emphasizes the trust into their work.

Which of these theories is valid? In many corporate vision statements and keynote speeches, theory Y is conjured, while at the same time, the actual daily working environment is based on X. This has always been that way! I have written several books that point out that people with a dominating left brain hemisphere tend to favor theory X, while most of those with a dominating right side of the brain keep up theory Y. With a dominant right brain hemisphere, I myself need to accept many statistics that show where the majorities are: Approximately 80% of all humans as well as most managers are left-brainers. About two thirds of all software developers are right-brainers. I believe that exactly here, in the middle of those two brain hemispheres, the root cause of the controversial discussion between structure and flexibility can be found. The question whether projects should be managed in an agile or traditional style touches the deepest beliefs of the involved. Looking at the majorities from a pure statistical point of view, it becomes evident that the executive power and the creation of innovation are on different sides of the game. I have already talked about these ideas in front of many disbelieving faces and I know what you might be thinking while reading these lines.

Look out how often the principles of agile software development highlight the value of each individual, self-responsibility, and team spirit! As a contrast, compare how often control, plans, schedules, budgets, and milestones take leading roles within traditional methods. Do you understand how deep this war of ideologies is? Since I am convinced that it is really a war, it cannot be overcome easily. I am afraid that even the compelling success of agile programming will probably not be sufficient!

I have already proved to many many managers that innovation does not emerge from planning and the administration offices, but instead from artist colonies like Silicon Valley. However, I strictly need to avoid the term "artist," or I would lose the debate immediately.

Perhaps agile development might flourish in places where it gets the opportunity to flourish – with an executive sponsor, who is personally convinced of theory Y …. And the odds are good that this will work out fine!

You can read this appealing book, an easy read about agility, like a journey into a world in which developers can work how they really want to work: acting meaningful and as humans. However, you also need to understand with all respect that traditional managers will stick to their numbers, dates, and plans with their entire left-brained heart. They will claim as well to be treated as humans with their very own patterns of behaving and ways of thinking. It is really a very delicate question: Should work be done according to the gusto of a structured management? Or should it be done according to the programmer's attitude? Should structure or sense drive a project forward? I am certain that we will begin to debate again immediately – X or Y? Are developers our top experts or skilled resources?

The authors of this book are calling for a cultural change and an altered way of thinking. They place their bets on other management styles (Y) and "evangelize" the readers as they put their vision into words. They have managed to create a wonderful emphatic pleading for inspiring projects. The authors also show you successful and large real-life projects from the same company I am working for. Sure, IBM tends toward structure – at least in general, but those who firmly want it different will only find the most minimalist head wind of numbers and planning blowing into their face, combined with an appreciating smile.

Mannheim, June 2009   Prof. Dr. Gunter Dueck,
                                                                      Chief Technologist,
                                                             IBM Innovation Network
                                            IBM Distinguished Engineer and Member
                                                  of the IBM Academy of Technology
                                                                         IBM Germany

About this Book

Economic pressures are tough for virtually anyone in the IT industry. Accelerating time-to-market, cutting development costs, or dealing with painful constraints in budgets and staffing are just a few challenges to highlight in this context. Furthermore, a development project needs to adjust to constantly changing requirements within a turbulent environment.

As a result, software development practices are going through a continuous evolution to make them more efficient and more flexible.

Agile software development is one buzzword in this context. Originally, most principles of agile development were derived from proven industry concepts like lean production. They were adjusted and amended to match the specific needs of creating high-quality software very efficiently. Today, manifold agile methodologies like Scrum, Lean Software Development, Test-Driven Development, Agile Unified Process, or Extreme Programming have been successfully introduced in serious and mission critical development activities.

But what is agility?

What kind of project is predestined for an agile development approach?

What are the limitations and opportunities?

What are the pitfalls and chances?

What are the challenges you need to master when you move towards agility?

Agile software development aims at the right balance between reliable structures and sufficient flexibility to accommodate change. Agile practices emphasize collaboration between humans as the source of customer value: let the people perform and excel without the confinement of rigid processes. However, most of all, agility is about lowering the center of gravity by letting the teams assume responsibility within a climate of trust and involvement.

This book will be your guide to agile software development.

We want to give a solid overview of agile development practices in a book that is easy to read and gives readers a fast and comprehensive start into their own agile project planning and execution. We show the advantages and challenges of agile software development and will outline the impact on development teams, project leaders, management, and software architects. We have summarized our own experiences as practitioners into a comprehensive set of considerations on project planning, software design, code development, and testing. This book will help to shape a project climate that can nourish collaboration and innovation by unleashing the power of teams.

The ideas and considerations within this book can be applied to any kind of software development activity or organization that wants to benefit from the merits of agility: software solutions within customer engagements, product development, research projects, small projects, start-up companies, and large development organizations pulling together people from many locations.

One particular focus is on moving large development projects from a traditional waterfall project approach towards agile practices, in order to become more flexible and improve the ability to react quickly to changing project constraints.

In this book, we share first-handed experiences gained when introducing agile software development in IBM's WebSphere Portal product development.

The Audience of this Book

The comprehensive and profound overview of agile software development makes this book very valuable for a wide audience of interested readers. Following the main thread of this book, they will find an easy and quick entry into the related topics.

Business managers and executives will learn what impact agile practices can have on their projects. They will see where agile software development can help businesses to develop software faster and more flexibly with regard to late-changing customer requests. We want to motivate the readers to rethink about organizational setup, leadership style, and management culture.

Those software architects and project leaders who need to react to the changing global environment, which forces software projects to adapt to changing requirements more quickly than ever, will read about how agile software development helps to live up to these new challenges. This book gives an overview of which methods are used today and how to apply them to a specific project. We will include practices to plan and monitor projects. We will also discuss the impact on architecture, leadership, and team setup.

Application developers getting involved with a project that is using agile software development methods for the first time will find a profound introduction to the subject. They will rapidly get up-to-speed with these new concepts and will understand how these can change their role within the project.

With this book we especially invite every individual development project team member to begin to evangelize agile thinking within their environment. Keep in mind that agile software development is not a fixed set of rules and processes set in stone by the project leader. It is rather a set of evolving ideas and thoughts, which are brought forward by practitioners who actively influence their project.

Just imagine – you can be the one who makes an agile project successful!

About the Content

We have broken this book into chunks that can be read in almost any sequence:

- *The Flaw in the Plan:* We set the stage by providing a background on why companies are starting to think about and develop agile practices. This chapter proposes five fundamental principles, which can serve as a main thread guiding red line through an adoption of agile software development.
- *Traditional Software Development:* In this chapter we outline the key characteristics of a traditional waterfall-oriented project management. This section serves as a contrast to agile software development and helps to better understand the differences.
- *Overview of Agile Software Development:* This chapter introduces the most common methods and approaches that make up agile software development.
- *Tooling:* In addition to the description of agile methodologies, this chapter provides a short overview of supporting tools.
- *Considerations on Teaming and Leadership:* In this chapter we focus on teaming structures, collaboration, management styles, and how the wisdom of the crowds can contribute to a project's success. We want to evangelize a change in an organization's culture and in the mindset of the participating stakeholders.
- *Considerations on Planning and Architecture:* Here we discuss what needs to be considered to start an agile project. This chapter includes considerations on project management and architecture and aims at simplifying processes as well as finding the right balance between structure and flexibility.
- *Considerations on Project Execution:* This chapter covers the execution part of an agile project. This includes the fundamental ideas of small, time-boxed iterations and continuous integration. We will share considerations on measuring progress, understanding the current quality, and reacting to change.
- *Mix and Match:* Individual projects have different needs and can choose those agile practices that suit them best. In this chapter we look at projects moving from the waterfall model to software development. We will give a real-life example and show how WebSphere Portal has adopted agility.

- *Summary and End:* Last but not least, we highlight the key characteristics of
  projects that have successfully adopted agile software development. In this
  chapter, we also wrap up the wide set of considerations, experiences, and best
  practices.

About the Authors

Dr. Thomas Stober received a master degree from the University of Karlsruhe and a Ph.D. from the University of Stuttgart, Germany. After 5 years of research at the Fraunhofer-Institute IPA, where he focused on mobile computing and information logistics, he joined IBM's Pervasive Computing Division in 1998. As a technical leader and architect, Thomas developed smart card technology, data synchronization solutions, and was a member of several related standardization activities. During the last couple of years, he has been working for Lotus in Westford, Massachusetts and Boeblingen, Germany.

Thomas is the lead architect for the WebSphere Portal Foundation and is a recognized expert on strategic topics like Internet Portals, Web 2.0, and enterprise collaboration. In particular, Thomas is the key driver for introducing agile development practices into the Portal organization.

Thomas publishes articles and books and presents his work regularly on conferences and at university lectures. Together with Uwe, he has published the Pervasive Computing Handbook.

Uwe Hansmann is currently the Release Manager for WebSphere Portal. Uwe has 20 years of experience in software development. He joined IBM in 1993 as a developer using object-oriented concepts development software running on OS/2. He soon led various projects and teams within IBM's Software Group on topics like Smart Cards, Pervasive Computing, or synchronization. During the last couple of years, he led large product development projects at IBM that used agile software development methods like the development and release of WebSphere Portal 6.1.

He has also works directly and very closely with some major customers and was the Secretary of the Open Services Gateway Initiative. Uwe received a Master of Science from the University of Applied Studies of Stuttgart in 1993 and an MBA from the University of Hagen in 1998. He is a PMI-certified Project Management Professional (PMP) as well as an IBM-certified Executive Project Manager.

Uwe is also the co-author of Smart Card Application Development Using Java, as well as SyncML – Synchronizing and Managing Your Mobile Data.

Acknowledgements

We have had the unique opportunity to accompany the WebSphere Portal product from its very beginnings to the present day, when the product has become the market-leading internet portal product. We have witnessed how the product grew over the years and how difficult the move to an efficient agile development approach was. Teaching an elephant to dance has been a substantial endeavor, in which many evangelists have participated with their heart's blood and passion. We thank these individuals for their support and persistence and for all the intensive discussions: Inge Buecker, Doug Geiger, Bill Krebs, Howard Krovetz, Sue McKinney, Dave Ogle, Brian Ricker, Wolfgang Stürner, and Stefan Weigeldt. Without them, we could not have written this book.

We also thank the International Business Machines Corporation, and in particular, Lotus for having provided us with the opportunity to write this book.

Numerous people supported us with in-depth reviews of the book or provided us with their invaluable expertise. We are indebted to Erich Bayer, Reinhard Brosche, Doug Cox, Gunter Dueck, Rainer Dzierzon, Hermann Engesser, Gabriele Fischer, Doug Geiger, Dorothea Glaunsinger, Stefan Hepper, Athiappan Kumar, Marshall Lamb, Martin Scott Nicklous, Brian Ricker, Stephan Rieger, Wolfgang Stürner, and Dirk Wittkopp. We especially thank Elke Painke for the thorough review of the manuscript and her indispensable comments and suggestions.

Last but not least, we also thank Anna, Anne, Chia, Laura, Melanie, Michael, and Sandra for the borrowed time.

Böblingen, May 2009

Thomas Stober
Uwe Hansmann

Contents

Chapter 1
The Flaw in the Plan

1.1 The Delusive Perception of Having Anticipated Everything

Once upon a time, Gustav II Adolf, the king of Sweden gave orders to build the biggest war ship that would ever set sail on the oceans of his time: The Vasa. Among the planned key features of the vessel were two decks stuffed with cannons. Such a power of arms has been unprecedented. The king hired the best ship makers from all over Europe. An impressive gathering of skills, talent, and ambition took place in a shipyard near Stockholm, Sweden. The admiral of the fleet personally took over the project management. The construction was led by an architect from the Netherlands and the designated captain of the new ship. In those days, there were no construction blueprints or static calculations. The architect applied proportions of other existing ships and supervised the work of the craftsmen in the shipyard personally.

During a period of 2 full years, the team created the huge ship from 1,000 oak trees. The afterdeck raised high above the waterline and had been enriched with precious ornaments to honor the king: There were beautiful wooden statues resembling roman warriors. There were lions, mermaids, and images of Greek gods. The craftsmen attached the huge masts and loaded the 64 heavy cannons. The caliber of the cannons on the upper deck had been increased late in the project, because there had been rumors that the enemy was attempting to build a similar ship. Despite this significant change of the weight distribution, no replanning of the project occurred. The architect continued to apply the proven proportions to the Vasa's hulk, which have worked so well for all of his other ship constructions in the past.

At the very end of the construction phase, the architect ordered the final testing of the ship. There were two test cases scheduled: First, 30 men should run back and forth from one side of the ship to the other. Second, the number of men for that exercise should be doubled.

Unfortunately it turned out that the ship was almost out of control after the first test. The ballast close to the keel did not match the heavy load of cannons on

T. Stober and U. Hansmann, *Agile Software Development*,
DOI 10.1007/978-3-540-70832-2_1, © Springer-Verlag Berlin Heidelberg 2010

the upper deck. The heeling of the ship threatened. It turned out that the ship's body had been designed too slim and far too high.

Captain, architect, and admiral had to make a decision: The celebration event of the ship launch had already been scheduled and was only a few weeks ahead. The king urged them to send the new ship to war as quickly as possible. Finally, the architect agreed to ignore the issues and confirmed the launch to his majesty.

On August 10th, 1628, a colorful crowd of people gathered around the harbor of Stockholm. They wanted to witness how the heavily armed war ship Vasa set sail for the very first time. On board, the crew opened the portholes for the cannons to fire salute and greet the cheering citizens. It must have been an amazing view to see the 1,300 m^2 of canvas rising in front of Stockholm's skyline, as the ship steadily took up speed.

Suddenly a slight gust of wind turned the Vasa to the side. Water entered the hulk through the portholes immediately. The ship sank within minutes. The entire journey of the Vasa was approximately 1,300 m and it took less than half an hour. More than 30 sailors died during this accident.

Developing modern software faces challenges that are not that much different from building medieval war ships: There is the "unknown," since many technical problems cannot be understood in their entirety at the beginning of the project.

There is the complexity of the infrastructure on which IT solutions are built. There are challenges arising from international, distributed teams. There are commitments to stakeholders, as well as not yet identified dependencies, exploding costs, and rapidly approaching deadlines. There are requirements that come up late in the game and decisions that are hard to make. There are consequences that are hard to predict. There are always unexpected issues to mitigate. The required knowledge, applied set of skills, and available technical information to digest is soaring at a tremendous rate. And to make everything worse: The complexity of IT solutions can easily get out of control. At the same time the need for cost-effectiveness is growing mercilessly.

We are quickly getting into the dilemma that we want to understand all these difficult dependencies as the baseline for our planning activities, but the dependencies will not be fully known until the project is completed. The complexity is quickly exceeding the scope that is manageable and comprehensible by individuals.

We are reaching the limits of engineering and science: Since the days of Sir Isaac Newton (1643–1727), science has been founded on a strictly deterministic view of our world. Technical innovation deals with precisely modeling correlations between cause and impact.

Pierre Simon Laplace (1749–1827) proclaimed that our future could be determined, if the exact state and all rules of the universe were known.

> We may regard the present state of the universe as the effect of its past and the cause of its future. An intellect which at a certain moment would know all forces that set nature in motion, and all positions of all items of which nature is composed, if this intellect were also vast enough to submit these data to analysis, it would embrace in a single formula the movements of the greatest bodies of the universe and those of the tiniest atom; for such an

intellect nothing would be uncertain and the future just like the past would be present before its eyes. (Pierre-Simon Laplace [4])

It was Frederic Winslow Taylor (1856–1915) who took this traditional view of engineering and applied it to the management of organizations. He propagated the division of labor and suggested splitting the planning aspect of work from its execution in order to achieve higher productivity. Specialists reach perfection by extensive scientific study and analysis of the applied methods. Each single working task needs to be specified precisely without any scope for the individual worker to excel or think.

Scientific management requires first, a careful investigation of each of the many modifications of the same implement. The good points of several of them shall be unified in a single standard implementation, which will enable the workman to work faster and with greater ease than he could before. (Frederic Winslow Taylor [11])

This methodology was exercised extremely successfully throughout the twentieth century. It significantly influenced Henri Ford's first conveyer belts, and was applied to the manufacturing industry, to financial institutions, as well as to any kind of product development or project execution. You can attempt to cope with the increasing complexity of a project by perfecting your tools, processes, systems, and techniques to reflect even more data to be analyzed and parameters to be examined. Maybe the next version of your favorite Gantt chart tool will add a dozen new powerful features to make your schedule planning more effective. Maybe you will add more engineers to focus on even more precise planning exercises and quality forecasts.

But this approach does not scale forever, since especially software development projects usually face late challenges and significant unknowns. And it will most likely not improve your ability to deliver real customer value quickly, at low cost and with high quality.

Life is what happens, when you are busy making other plans (John Lennon)

Mathematicians already know the difficulties in solving non-linear functions with a lot of unknown parameters very well. Once you get started working with probabilities and approximations, you will be permanently kept busy with adjusting and ordering your models. In 1927, the physicist Werner Heisenberg made it very clear when he proclaimed his uncertainty principle: the values of two variables (for example the momentum and position) cannot both be determined with arbitrary precision at the same time. The more precise one property is, the less precise the other will be.

Another physicist, Jules Henri Poincaré (1854–1912) wrote:

A very small cause which escapes our notice determines a considerable effect that we cannot fail to see, and then we say that the effect is due to chance (Jules Henri Poincaré [9])

Even though computers allow the creation of much more precise models and predictions, there will always be limits and failures, as the quality of any model depends on the quality of the parameters you feed into the model. Once you cannot describe the actual state of a system, you will get uncertainty about your results.

We need to accept that the deterministic view of our world can no longer endure beyond a specific level of complexity.

All of this applies to software development projects that attempt to accomplish more than just "Hello World" sample code: A project plan for a large development endeavor that spans a period of several months and includes a lot of details takes a tremendous and extremely time-consuming effort to create. Despite all hard effort, such an elaborate plan will most of all give you the dangerous illusion of having the project execution predictably and well under control. In reality the plan will always be based on many assumptions and include a lot of rough guesswork.

According to Murphy's law anything that can go wrong will go wrong. The world keeps changing while the "perfect" project plan is being executed. Only a fool will blindly trust and follow a plan without questioning it every day. As we have seen looking at the Vasa, an additional deck of cannons can easily turn a project into a nightmare. A quote from Mark Twain nicely points out an important risk:

> It ain't what you don't know that gets you into trouble. It's what you know for sure that just ain't so. (Mark Twain)

A large project following a classical waterfall approach might define distinct milestones, a development phase, and a closing testing cycle. But in reality, these distinct phases will rarely be so distinct and isolated. Instead, there will be dependencies and corrective loops circulating back to earlier phases in order to adjust. Typically, unforeseen issues, such as a new requirement, a design flaw, a growing number of bugs beyond the expected level, or redirection of resources to other activities make it necessary to rework the plan over and over again. Communication

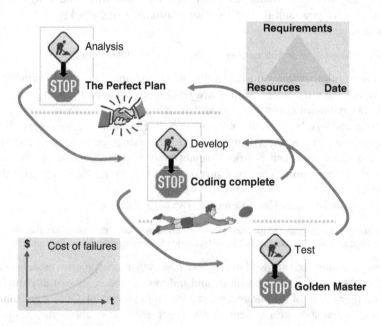

Fig. 1.1 Traditional waterfall approach

and interfaces between different organizational units are a challenge in large distributed teams with a centralized project management in between. Bringing independently developed pieces together in order to assemble a complex use case requires a significant integration effort before the overall system reaches a satisfying level of stability. After coding and unit testing is finished, there is often a handover to a different team that is responsible for function verification testing as well as for system verification testing (Fig. 1.1).

In many projects the problems become really, really pressing towards the end of the final test phase. The test team keeps finding more bugs than anticipated. Obviously the cost of fixing problems increases significantly the later an issue is detected. The accumulating bug fixing work exceeds the capacity of the developers, while the necessary turnaround time for the fixes causes further delays for the testers. Project management is quickly running out of options: at that stage of the project, content removal is not really an option, as the code is already done, although not stable. Delaying the shipment is not a good option, as customers do rely on the promised delivery dates. And sacrificing the quality is not acceptable either. It takes a tremendous, costly team effort to solve the situation and ensure that a solid product is still shipped on time. Doesn't all this remind you of the story of the Vasa? Or maybe even a project you have experienced yourself?

1.2 Accept the Uncertainty and Adapt

The problems which turn many projects into a nightmare have already been stated above:

It's the complexity of our technology, the worldwide work split, and especially the dynamics of market-driven needs, combined with the budget or the capacity which is available to deliver this project. It becomes extremely hard and expensive (if not impossible) to elaborate a precise long-term plan, which takes all these constantly changing constraints into account.

Let us reemphasize this important point: Even the most precisely elaborated project plan can only give the delusive perception of having anticipated everything, as fundamental assumptions can change any time.

It becomes evident that successful projects must be organized in an adaptive, flexible way in order to quickly react to the unforeseeable. While the conductor of an orchestra can rely on a well-composed score, a project leader must learn to accept that the final version of the score will never exist in real life, as there is always too much ambiguity and uncertainty.

Once you accept that a software development project is a substantial endeavor and that the unexpected happens, then you can prepare yourself to focus on the ability to adapt quickly to face any kind of forthcoming challenge.

If you accept that change is the only thing that will happen for sure, this will have impact on the chosen project management practices, organization structures, teaming setup, and leadership styles, as well as on the suitable development strategies.

One key aspect to consider within a turbulent environment is that any form of extensive centralized planning will eventually become a bottleneck and impede the progress. A key shift in the mentality is needed in order to let excessive control and steering go and move towards an attitude of adapting and evolving.

The Nobel Prize winner and economist Friedrich August von Hayek (1899–1992) suggested a "spontaneous order" as the solution to deal with the limitations of human cognition and articulation. Such an order, which is driven by autonomously acting people und doesn't follow any predefined master plan, will be superior to any formally structured order that has been set up and coordinated by the most genius mind. Spontaneous systems are subject to a continuous evolution, driven by the effect of events and coincidences. They optimize themselves very similarly to the way living organisms advance in evolution.

A good example is the free market economy, which lets people trade and shape markets in a free and open manner. The past century showed that this order has been much more successful in adapting and accumulating value than a centrally planned economy. We believe that this still holds true after the crash of the financial markets in 2008. It remains to be seen if the influence of added government control will affect the markets in the long run more significantly than the self-healing powers of the worldwide economy.

> Unhampered market economy produces outcomes that are the results of human action but not of human design (Friedrich August von Hayek [12])

In nature you can find many examples of assertive organisms that are incredibly capable of adapting to their environment. Species manage to find improved methods and tools to fulfill tasks that are advantageous for their survival and success. Improvements evolve via a combination of the continuous production of small, random changes, followed by natural selection of the variants which are best-suited for their environment. Improvements can be behavioral or physiological. And there is an incredible diversity of possible solutions for a given problem.

One example for spontaneous order is an ant colony. The queen does not give direct orders and does not instruct the ants what to do. Instead, ants are autonomous local units, which communicate with their peers based on the scent they are emitting. An ant that finds food marks the shortest trail by applying a chemical signal for others to follow. Despite the lack of centralized decision making, ant colonies exhibit complex behavior, like learning: if an obstacle blocks the trail, the foragers will add trail markings to the detour. Successful trails are followed by more ants, reinforcing the scent on better routes and gradually finding the best path. Once the food source is exhausted, no new trails are marked and the scent dissipates. Different chemical signals are used to exchange information, like the task of a particular ant (e.g., foraging or nest maintenance). Any ant without a scent matching that of the colony will be attacked as an intruder.

To give another example: rats have managed to become immune to poison. The population's DNA had mutated so that blood coagulation is not affected by the poison.

Today, resistant brown rats exist in several parts of Europe as well as in North America. This mutation has not been triggered by intelligence or innovative planning. It has been a sheer coincidence which has been made possible because of the incredible speed of their reproduction.

The ability to transform very quickly is essential for survival.

> It is not the strongest of the species that survives, nor the most intelligent; it is the one that is most adaptable to change. (Charles Darwin)

If you translate all of this into the language of project management, you will learn that change cannot be accomplished fast enough if the organization is driven strictly from the top to bottom following a master plan. The abstraction level of a centralized project-wide view will imply delays in recognizing issues, and further delays until effects of any actions become noticeable. While the overarching strategic direction of a company or a project needs to be defined to a large extent from the top to the bottom, the detailed planning and execution should be directed in a decentralized, bottom-up approach. We need to understand that activities and structures within the individual teams require a high level of autonomy and vitality to live up to the imposed requirements.

Turnaround speed and pace are further accelerated when the product is optimized and developed in many short iterations, instead of following a long-term schedule, which would define several fixed target milestones many months ahead. Don't waste time on future details which will need to change over time anyway.

Iterations will help to structure a comprehensive project into smaller manageable units. Going forward with short iterations rather than planning for the future in advance will allow focusing on the present and continuously adapting to incoming requirements. Focusing on the present also means to continuously produce small and valuable incremental changes to a product. Each iteration will go through a full development cycle including planning, design, coding, and test. Throughout the project, prompt prototyping efforts will allow to choose the best of breed and make fact-based decisions as teams make progress.

We have not yet addressed the probably most important question: Why does adaptation happen and what is the driving force behind it? Or you might as well ask: What is the purpose of life?

In the context of evolution the answer is pretty obvious: surviving and reproducing is what makes a species successful. In the world of business, the overarching goal will be to make profit by creating customer value. There will be more fine-grained amendments added to that oversimplified purpose by each individual company or project. In an agile organization, which optimizes its capability to adapt, there will be a set of high-level goals that reflect long term strategy and tactical directions. Derived from these, there will be more detailed goals that serve as guidelines that drive the teams to contribute to the greater good. They need to be transparent, consistent, and well understood. Once the goals are agreed on, the individual teams will be able to organize themselves and operate autonomously to a large extent within their given scope of work. Part of the responsibility and authority of each team is to translate their goals into specific requirements and tasks to implement.

1.3 Involving the Teams

Key competence of a leader in such a changing environment like our IT industry is the ability to establish an open project structure in which responsibility is shared. Innovative and efficient teams are not managed in a centralized top-down management approach. It's all about getting the teams involved as stakeholders, rather than as suppliers.

Management by objectives, as described above, is capable of bringing teams into line even without a need for ubiquitous micromanagement: While the goals outline the overall strategy and the directions from top to bottom of an organization, a high-performance culture needs to encourage a strong influence and grassroots initiatives emerging from the base as well. Lowering the center of gravity by inviting teams to act as entrepreneurs will help to increase the speed of decisions. It will also tap into the unfiltered pool of available talent and skills.

Successful organizations will become highly dynamic organisms that are built from independent, self-organizing units that collaborate towards a common set of goals.

To make one point very clear: When we identify centralized planning as a limiting factor for vitality and innovation, we are absolutely *not* implying that leadership will become less important. On the contrary:

Leadership by defining strategic goals with an appropriate level of abstraction, rather then following the temptations of micromanagement, requires a distinct visionary talent. Only the clear and unambiguous communication of the project's (or company's) strategy and direction will ensure that all teams jointly move successfully toward a common greater good. Management's job is to eliminate impediments. Leaders need to inspire others and disseminate enthusiasm about the common goals. The better the personal goal of each individual is aligned with the corporate goals, the more powerful the performance of the entire team will be.

Regardless of the decided goals and strategies: Leaders need to also boost the talent, skills, and creativity that exist in their teams. They need to foster a climate of participation, which turns every single individual of the team into a stakeholder who assumes responsibility and contributes to the overall success with his or her very unique skills. Rather than being a classical orchestra with its deterministic score, a software development project is like a Jazz session in which conductor and team mix outlined structures and patterns with improvisation and interaction.

1.4 In Search of Structure

We want to live a software development project like a Jazz session and mix structure with improvisation. But how? How can we avoid improvisation turning into anarchy and chaos? If there is no generic master plan, and if we condemn too much structure, are there any guidelines to follow?

Again, we would like to borrow some concepts from mathematics:

Benoît Mandelbrot discovered amazing geometrical shapes by repeatedly applying very simple mathematical equations. The created objects resemble rough

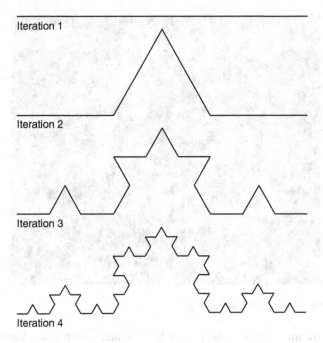

Iteration 1

Iteration 2

Iteration 3

Iteration 4

Fig. 1.2 The "Koch curve"

landscapes, mazy patterns, and bizarre figures, which pretend to be irregular and not predictable. Mandelbrot named these fragmented object "fractals" and established the world of fractal mathematics.

Fractals overcome shortcomings of the traditional geometry influenced by Euclid, which limits its focus on ideal deterministic forms and is not able to describe complex irregular shapes as they exist in real life, like a coastline, a mountain, or a cloud.

> The world that we live in is not naturally smooth-edged. The real world has been fashioned with rough edges. And yet, we have accepted a geometry that only describes shapes rarely – if ever – found in the real world. The geometry of Euclid describes ideal shapes – the sphere, the circle, the cube. Now these shapes do occur in our lives, but they are mostly man-made and not nature made. [. . .]I love Euclidean geometry, but it is quite clear that it does not give a reasonable presentation of the world. Mountains are not cones, clouds are not spheres, trees are not cylinders, neither does lightning travel in a straight line. Almost everything around us is non-Euclidean. [. . .]I conceived, developed and applied in many areas a new geometry of nature, which finds order in chaotic shapes and processes. Fractal geometry is the geometry of irregular shapes we find in nature, and in general fractals are characterized by infinite detail, infinite length and the absence of smoothness (Benoit Mandelbrot [6])

The "Koch Curve" in Fig. 1.2 belongs to the earliest examples of fractal curves that have been described. Its overarching mathematical rule is to start with a line segment and iteratively apply the following transformation:

• Take each line segment of the Koch curve and remove the middle third.

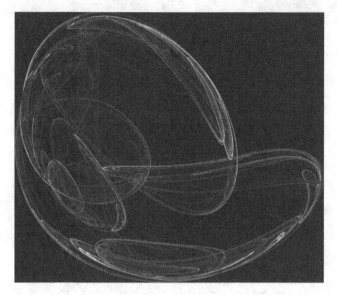

Fig. 1.3 An instance of a "Fractal flame"

- Replace the middle third with two new line segments, each with a length equal to the removed part, forming an equilateral triangle with no base.

The Fractal Flame in Fig. 1.3 is another example. Rules for constructing this fractal are based on non-linear functions generating esthetically pleasing images [3]. Fractal Flames are applied commercially for instance in Adobe's "After Effects" video editing software.

Although the irregular and fragmented shapes of the fractal geometry appear to be extremely erratic at the first glance, each fractal follows a set of simple, fundamental principles:

- A fractal is too irregular and complex to be easily described by traditional Euclidean geometry.
- A geometric shape can be split into parts, and each resulting fractal is a reduced-size copy and variation of the whole. All fractals are self-similar, which means that they have a similar structure. Despite the fact that each individual instance is unique, they organize themselves while following a common scheme and purpose (which is somehow a kind of goal expressed by a mathematical formula).
- Fractals are vital. This implies a feedback mechanism based on recursion, which drives them to change shape and adapt to changed parameters in every new iteration.

Fractals can be used to model and study complex real-world objects, for which Euclid's geometry will fail. There are many application examples, like modeling of dynamical systems, describing plant growth or economy, as well as dealing with data compression or computer-generated imagery (CGI) in movies. Just to name a few.

Fractal geometry is a new language. Once you are able to speak it, you can describe the shape of a cloud as precisely as an architect can describe a house. (Michael F. Barnsley [1])

Within his vision of a fractal company, Hans-Jürgen Warnecke has applied principles of fractal mathematics by suggesting autonomously acting units within a flexible organizational structure in a manufacturing plant.

The creation of manageable units within an enterprise is a promising approach to act fast and close to the market needs. Fractals provide services, they are subject to continuous change and are integrated into a process of establishing goals. They organize and administer themselves. They navigate. (Hans-Jürgen Warnecke [13])

What will happen if we dare to transfer these fundamental principles to organizing a software development project? "Act fast," "continuous change," "goals," and "organize themselves": aren't these exactly the properties we want to bring into a simple structure?

Let us consider an autonomously acting team of a software development project as a fractal unit! By applying and extending the principles of fractal mathematics to such a development fractal, the resulting characteristics can be:

- *Self-similarity*: All fractals are alike: The purpose of each fractal is to create customer value and implement a corporate strategy. A corporate culture sets common values and rules of engagement, e.g., a climate of trust. Fractals have a holistic, interdisciplinary view of the deliverables they are working on. Each fractal is pursuing end-to-end use cases, which can span multiple functional areas and can involve different kinds of skills. Being self-similar does not imply that two different fractals that happen to have the same goals will be identical: Each fractal might organize itself entirely differently and apply a different approach to pursue its mission.
- *Goal-orientation*: The fractal is pursuing a set of explicitly described goals, which are derived from the project's high-level goals and an overarching corporate strategy. Defining goals mixes a top-down and a bottom-up approach: While high-level goals are a given to a large extent, each fractal will translate them bottom-up into precise goals based on their best judgment and within their given scope of action. A common set of high-level goals and an overall strategy is essential to make sure that all individual fractals are lined up properly and are heading in the same direction.
- *Self-organization*: A fractal is involved in its formation, variation, and disbandment. As a team, a fractal will have sufficient responsibility to fulfill its mission. Team members are empowered to organize their work in a way that promises the best possible results. This can include technical aspects (e.g., choice of tools), organizational aspects (e.g., ways of customer involvement) or social aspects (e.g., communication and roles).
- *Self-improvement*: A fractal is continuously responding to changing constraints, such as customer requirements, design decisions, actual progress, quality level, available team size, and constraining project dates. A fractal will adapt and evolve both its organizational structure and its deliverables.

- *Vitality*: Each fractal is making progress in short, recurring iterations with small increments to the product. To maintain vitality, a fractal should focus its planning on the present. A fractal decides as late as possible on the final content it will eventually deliver, while usable and tested pieces of functionality are made available as early as feasible. Frequent interaction with customers will allow for prompt quality feedback and boost the speed of adaptation and continuous improvements. Fractals are embedded in an organization with a flat hierarchy, which supports quick interaction, selective competition and inspiring cooperation between each fractal.

Adopting these five fundamental principles of fractals will provide some simple and crisp guidelines for leadership, teaming, and an organizational setup of a development project. They ought to encourage you to live your project like an agile Jazz session and find the right balance between structure and improvisation. Vitality is the underlying force for driving self-organization and self-improvement. Flexibility of decisions and plans allows for quick adaptability to changing constraints. Being goal-oriented ensures that the results of the fractal will focus on customer value and align with the greater good of the overall project. Teams will be empowered to perform within a climate of trust, participation, and intensive collaboration.

1.5 Agile Software Development

How do all these considerations relate to agile software development and the scope of this book? Well, agility is, most of all, a perfect synonym for the kind of adaptive and lightweight development approach we kept evangelizing throughout this chap. While the five principles above nicely summarize key characteristics, the ideas of agile software development are adding more thoughts to its specific implementation:

It is important to emphasize that agile software development is not just a single set of rules to follow in order to be successful. Agile methodologies comprise leadership concepts, lightweight project management processes, engineering best practices, development techniques, as well as supportive tools. They address the entire enterprise on all levels and in all disciplines.

An agile approach *grows and evolves* over time by *mixing and matching* ideas contributed by *practitioners*.

This summarizes a set of important truths:

- *Grow and evolve*: An agile project will not pick a formal process from the shelf and simply execute it. The actual process will shape out while teams develop hands-on and learn from the experiences being made. The development approach is subject to the same need for change and constant improvement as the developed code itself. When development processes mature from primitive to perfection, it can imply that we grow a notable complexity. But if a development process becomes too complex, it will die under its own weight. Therefore, a clear aim is to evolve towards simplicity instead: Simplicity with the right balance between structure and flexibility is the key to success!

- *Mix and match:* Agility is a pool of thoughts, out of which individual projects can combine their specific implementation of an agile development process. The team's way of working is a composition of practices, instead of applying a generic "one-size-fits-all" process. The resulting process will be tailored to a team's needs and context. Even within the same organization, different teams can adopt different practices.
- *Practitioners:* While agility will definitely need active sponsorship and support by top management, the actual implementation will be shaped bottom-up by the teams themselves. The applied methodologies are strongly influenced by practitioners, who disseminate their ideas into their daily working environment. This is quite different to classical development processes which are defined by specialists and are introduced within an organization in a top-down approach as generally binding. Driving the development approach from the bottom will help to avoid gaps between the defined processes and reflect what the teams are actually doing.

In Chap. 3 of this book we will cover the most relevant methodologies that have established themselves under the wide umbrella of agile software development, in much more detail.

Most of them promote iterative development in small increments and have a strong focus on teamwork. Plan and design of the project is outlined only at a high level, while the current iteration is detailed further. Teams are decentralized as self-organizing "fractals" and cover end-to-end functionality including test. They are equipped with the necessary authority to pursue their goals, rather than being micromanaged, thoroughly tracked or permanently inspected. In an agile environment, working and tested code is the preferred way to measure progress.

For now, here are a few well-known examples of agile thoughts to start with:

- *Lean Software Development* is a concept which evolved from the ideas of lean manufacturing in automotive industry. It includes thoughts on eliminating waste, adding customer value, and empowering workers.
- *Scrum* is an extremely efficient and streamlined process of managing and tracking teams.
- The *Agile Unified Process* combines a process with tools and is a derivate of Rational's Unified Process.
- *Test Driven Development* emphasizes the definition of test cases even prior to the actual implementation of the code.
- *Extreme Programming* is a toolbox of engineering practices like pair programming, or continuous integration.

All these methods and techniques have many things in common and are related. In 2001, a group of experts in agile software development published a set of guidelines as a common denominator of the methodology. This "Agile Manifesto" emphasizes the following four best practices to uncover better ways of developing software:

Individuals and interactions over processes and tools
Working software over comprehensive documentation
Customer collaboration over contract negotiation
Responding to change over following a plan
(Agile Manifesto [7])

Further Readings

1. Barnsley, Michael: Fractals Everywhere. CA, USA: Morgan Kaufmann 2000
2. Briggs, John: Fractals - The Patterns of Chaos. London: Thames and Hudson 1992
3. Draves, Scott; Reckase, Erik: The Fractal Flame Algorithm: 2008
   http://flam3.com/flame.pdf
4. Laplace, Pierre Simon: A Philosophical Essay on Probabilities. New York: Dover Publications 1951
5. Lesmoir-Gordon, Nigel: Introducing Fractal Geometry. USA: Totem Books 2006
6. Mandelbrot, Benoît: The Fractal Geometry of Nature. New York: W. H. Freeman and Co., 1982.
7. Manifesto for Agile Software Development.
   http://agilemanifesto.org/
8. Petroski, Henry: The Evolution of Useful Things: How Everyday Artifacts-From Forks and Pins to Paper Clips and Zippers-Came to be as They are. New York: First Vintage Books 1992
9. Poincaré, Henri: Science and Method. New York: Dover Publications 1914
10. Pór, George: Blog of Collective Intelligence.
    http://www.community-intelligence.com/blogs/public/
11. Taylor, Frederick W: Scientific Management - Comprising Shop Management, The principles of Scientific Management and Testimony before the Special House Committee: Harper and Row, 1964
12. Von Hayek, Friedrich August: The Road to Serfdom: Text and Documents: Text and Documents. Chicago: Chicago Press 2007
13. Warnecke, Hans-Jürgen: Die Fraktale Fabrik. Revolution der Unternehmenskultur. Heidelberg: Springer 1996
14. Womack, James; Jones, Daniel; Roos, Daniel: The Machine That Changed the World: The Story of Lean Production - Toyota's Secret Weapon in the Global Car Wars That Is Now Revolutionizing World Industry. New York: Rawson 1990

Chapter 2
Traditional Software Development

2.1 History of Project Management

Large projects from the past must already have had some sort of project management, such the Pyramid of Giza or Pyramid of Cheops, which were built more than
2,500 years BC. We cannot believe that those types of projects were possible
without some sort of project management. But at least we are not aware of the
project management practices that were used back then.

The 1950s mark the beginning of project management in the modern sense.
Before that, projects were already using Gantt charts (developed by Charles Gantt),
but during the 1950s more formal project management techniques were developed,
documented, and made available to the community (Fig. 2.1).

PERT, Program Evaluation and Review Technique, one of the first project
management scheduling techniques, was developed by Booz Allen Hamilton Inc.
as part of the Polaris submarine missile project. It was developed to schedule and
coordinate large projects. What you see in Fig. 2.2 is a CPM Network chart created
with Microsoft Project. In this flavor the nodes represent tasks; in the original PERT
chart the nodes represent events or milestones. Around the same time, the Critical
Path Method (CPM) was developed at DuPont, originally to plan the shutdown and
start-up of chemical factories before and after maintenance. CPM is very similar to
PERT. In CPM, the nodes represent activities or events (an event is an activity with
duration of zero) and dependencies are represented by arrows between two nodes.
Using Earliest Start- and Finish-time as well as Latest Start- and Finish-Time, one
can calculate the critical path in a CPM network diagram.

As project management was becoming more and more important in the 1960s,
the Project Management Institute (www.pmi.org) was founded in 1969 to advance
the practice, science, and profession of Project Management. The PMI offers
certification as certified project manager, which is currently the most widely
recognized project management certification in the industry. The Project Management Institute also publishes the PMBOK Guide, which contains the basic concepts
and techniques for traditional project management.

T. Stober and U. Hansmann, *Agile Software Development*,
DOI 10.1007/978-3-540-70832-2_2, © Springer-Verlag Berlin Heidelberg 2010

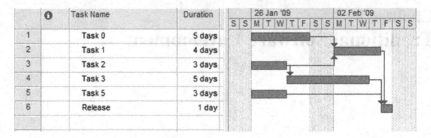

Fig. 2.1 Gantt chart

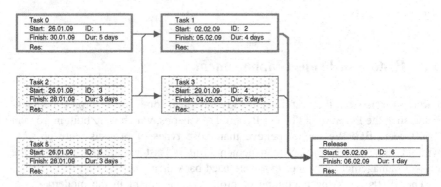

Fig. 2.2 Simple CPM network chart

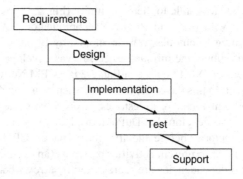

Fig. 2.3 Traditional waterfall approach

2.2 Waterfall Approach

The waterfall approach is the traditional approach used in smaller and larger projects for the last few decades (Fig. 2.3).

In the waterfall approach, each phase is completely finished before the next ones starts. In the next few chap. we will describe and discuss each phase in more detail.

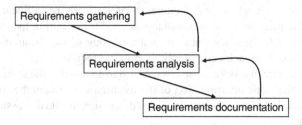

Fig. 2.4 The three steps of the requirement phase

2.2.1 Requirements

Every project starts with the requirement phase, during which all the requirements are collected, documented, and discussed with all the stakeholders. Stakeholders are all individuals or groups of individuals who have an interest in the project or its outcome. This is of course the project sponsor who is funding and promoting the project as well as the future users of the system. But there can also be others, such as the workers council, competing activities or interest groups in the company, the purchase department and so on. A stakeholder can also be a group that has a negative attitude against the project. It is as important that all stakeholders are identified and their requirements are collected, analyzed, and taken into consideration. All communication to all stakeholders must be managed in the best interest of the project.

Stakeholder management is usually divided into the following two activities:

- *Stakeholder analysis* is the phase during which all the stakeholders are identified and their interests in the project analyzed and documented. The team needs to poll all the stakeholders to get their requirements.
- A communication plan is put together during *stakeholder planning* to ensure that all stakeholders are kept involved, depending on their interest and needs.

The requirement phase contains the three steps shown in Fig. 2.4. During the requirements gathering steps, the team identifies all the stakeholders (also described above under stakeholder analysis) and interviews them to collect their requirements. The team then analyzes these requirements and documents them in the *systems requirements specification*.

The requirements can be separated into two major groups:

- Functional requirements, which are often described as use cases describe a user scenario or interact with the solution, such as "A user wants to calculate todays value of his or her stock trading account" or "As an administrator, I would like to change the default settings for new accounts." These use cases are then brought to a more detailed level during the Design phase.
- Non-functional requirements, such as the software and hardware environment that it should support, but also the performance characteristics the solution needs to meet.

The requirements analysis and documentation is ideally not a linear process, as the team should meet with the stakeholders several times during the requirements phase and present to them their current understanding of the requirements for the solution to collect feedback and potentially missing requirements.

Prototypes or screen mockups are often useful to ensure that the stakeholders and the team have the same interpretation of the requirements. Often the stakeholders' requirements evolve during the requirements discussion, as the discussion becomes more concrete.

Especially during the waterfall approach it is important that the requirements are completely captured and agreed on before the team moves to the design phase, as every new or changing requirement means that the team needs to go back to the requirements phase, which may invalidate work that was already done in an earlier phase.

2.2.2   Design Phase

During this phase the team creates a detailed design for the complete system as well as for each of the individual components. This is done at a level that developers can directly translate into code in the next phase.

2.2.2.1   Use Cases

The high-level use cases from the requirements phase are taken and detailed. Use cases describe the behavior and interaction of the system with a user or another system. It may depend on the complexity of the overall system how detailed a use case should be, but in general there should be a separate use case for each interaction with the system.

Use cases are not only used in traditional software development approaches but also in agile ones, they are called user stories in agile terms. Especially in agile approaches it becomes very important that user stories are on the one hand very granular and on the other hand also complete in the sense that the team could decide not to implement certain user stories without impacting other user stories.

A feature can be described in one high-level use case and later on broken down into several individual ones (Fig. 2.5).

There are a lot of different templates for use cases available and each team should verify which one fits best for its project.

Some sections are common for most of the templates:

- *Name:* The name should describe, in two or three words, what the use case is about, like "Update userid & password."

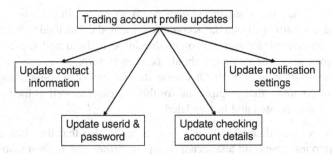

Fig. 2.5 A high-level use case broken into several more detailed ones

- *Number:* Often use cases are numbered within a particular numbering scheme, which should also show to which part of the system the use case belongs. For example, the use case *update contact information* could have the number 2-4-5, where 2 indicates that it belongs to the Internet Trading Front End of the system, 2-4 indicates that within the Internet Trading Front End this use case belongs to the high-level use case *trading account profile updates*.
- *Version:* Some templates suggest keeping a Version number to be able to identify the current as well as previous versions of this use case. Usually it is a good idea to also keep older versions of the use case around, so one could go back and see what was changed when.
- *Goal & summary:* The goal is a short summary of what this use case should achieve, like "user is able to update the password."
- *Actors:* Actors are individuals acting with the system, like an end user, an administrator, a clerk, or an investment banker, but can also be another system or a device, like an account statement printer.
- *Preconditions:* The state the system needs to be in before the use case can be executed, like, for example, the user needs to be authenticated or there must be at least one transaction since the last time the account statement was printed.
- *Triggers:* Triggers are conditions that cause the use case to be executed, like, for example, in the case of an account statement, that 30 days have passed since the last account statement was printed, or the user is requesting the print of an account statement. The preconditions need to be met before the use case is executed.
- *Main Flow:* The main flow describes the primary scenario of the use case, this is often also called the default case where the use case is executed without any error conditions, exceptions, or alternate selections.
- *Alternate flow:* Every path that the use case can take that is not the main flow described above. This can, for example, be if a user chooses to select something other than the default selections, such as that he or she wants the account statements mailed to the home address instead of viewing it online. Depending on the number of selections a user can make or the number of input streams from other devices or systems, there could be quite a number of alternate flows. Exceptions and error conditions can also be described here, but it may also make sense to put them in a section on their own.

- *Post conditions:* The state the system needs to be in after the use case was executed. Like after printing the account statement, the data field to keep the date for the last printed account statement should be updated and the counter for transactions since last statement should be reset to zero.
- *Comments, author, and date:* Of course the use case can contain any further remark or comment that is important for this use case, as well as the author and the date it was created and last updated.

In our view, a nice advantage of using use cases is also that they can easily be translated into test cases and test scenarios and therefore give the test team an easy entry into understanding the project and what the solution should do. In approaches with a test team that is separate from the development team and where the development team is not producing detailed use cases, it is often very hard for the test team to define the test scenarios. They then often have to re-engineer what the architects, software designers, and developers wanted to achieve by implementing a particular function. This can create a lot of frustration with the test team, but also with the developers, as the developers need to educate the test team and wonder why the test team doesn't already know what a particular function is supposed to do.

The relationship between different use cases and different actors is usually documented using use case diagrams. The use case diagram contains of course the use cases including the sequence of actions. It also contains the actor interactions with the use cases, in our example in Fig. 2.6 these are the call center representative as well as the customer. The associations between the actors and the use cases are represented by solid lines. In the example below you can see that the customer can only interact with two of the use cases and that the call center can interact with all four use cases.

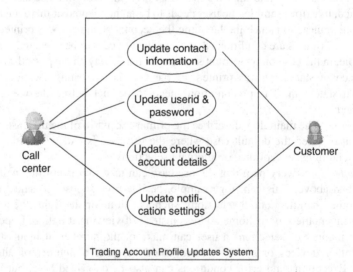

Fig. 2.6 Use case diagram

The rectangular box around the use cases represents the system boundaries to indicate the scope of the system.

The use case diagram is also part of the Object Management Group's (OMG) Unified Modeling Language (UML).

2.2.2.2 UML

The Unified Modeling Language (UML) is a popular way to model the system using object-oriented methods. UML is the evolution of and based on Rumbaugh's Object Modeling Technique (OMT) [14], the Booch method developed by Grady Booch [1], and Ivar Jacobson's Object-Oriented Software Engineering (OOSE) method [6].

From 1995 on Booth, Jacobson, and Rumbaugh all worked for Rational Software (now a division of IBM), Rational asked them to jointly work on a modeling language that would drive the further adoption of object-oriented design.

The UML 1.0 specification was adopted by the OMG in 1997 and version 2.0 became available in 2005.

UML provides 13 diagrams to model a software system, which are organized in three major groups (Fig. 2.7):

- Structure diagrams: class diagram, object diagram, component diagram, composite structure diagram, package diagram, and deployment diagram.
- Behavior diagrams: use case diagram, activity diagram, and state machine diagram.
- Interaction diagrams: sequence diagram, communication diagram, timing diagram, and interaction overview diagram. The interaction diagrams are derived from the behavior diagrams.

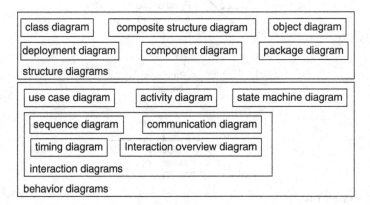

Fig. 2.7 The different UML diagrams and their relationships

Several tools automatically generate class definitions and code fragments based on UML diagrams.

2.2.2.3 Flowcharts

Flowcharts have been around for decades and were used long before object-oriented design was invented. They are especially useful for procedural languages, like Basic or C, and can be used to design the flows of complex systems (Fig. 2.8).

Flow charts are built using the following basic elements:

- *Start and end*: Each flow chart has a start element, indicating the start of the process as well as an end element, indicating the end of process flow.
- *Processing steps:* A rectangle is used to represent each step where something is processed, like a value increased or two numbers multiplied.
- *Arrows:* The flow and the direction of the process from one step to another is indicated using an arrow.
- *Decision points:* A rhombus (a unilateral parallelogram) is used to indicate decision points. From a decision point, there are usually two arrows leading to the next step, one if the result of the check is true, and another one if it is false.
- *Input/Output:* A parallelogram is used to represent read and write operations. This may be input from the user, from another application, or from another device.

Developers can then directly translate the processing flow into code.

The project moves to the implementation phase to code the design as documented after the complete design is finished, reviewed, and signed off by the team as well as the stakeholders.

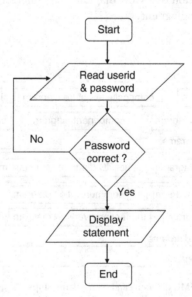

Fig. 2.8 A simple flowchart processing a password check

Fig. 2.9 Usual workflow
within the coding phase of a
waterfall project. The dashed
line back from "Integrate" to
"Write code" should not be
needed in a perfect waterfall
project

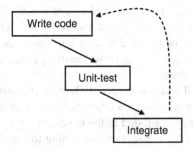

2.2.3 Implementation

Now that the design is finished and perfect (at least it needs to be viewed as perfect),
it is translated into code. Developers take the flowcharts, UML diagrams, and the
other design documents and translate them into code, component by component or
object by object. Each component is unit-tested on its own and a code review should
be done. Usually the integration is done at the end of the coding phase or at the
beginning of the test phase. This is often the time for surprises, as things may not fit
or work together as planned, and the team may have to go back to the design phase
to make the appropriate changes.

Depending on how tough the schedule was, the project may get in trouble right at
this point, as the plan was to integrate the components to form a perfectly working
system. But now there is a delay due to the integration issues (Fig. 2.9).

The risk of a big surprise at the end can be reduced by introducing different
milestones during the coding phase, at which point in time the system is integrated
and needs to provide a certain level of functionality. These milestones should
ideally be followed by a set of tests to verify that the functionality works correctly.

2.2.4 Testing

After the developers have declared that they are done with the coding phase, what is
often also called DCUT (Design Code Unit Test), the different units and components are put together into an integrated system. As already mentioned in the
previous chap., this is usually the time where the project goes south due to many
surprises and integration issues. This could be because developers have made
mistakes during coding, or that a component A behaves slightly differently than
another component B expected it to and now they simply don't work together.

Even with a waterfall approach is it usually good to have an integration phase at
the end of the coding phase or at the beginning of the testing phase. This integration
phase should be owned by the development team, which needs to prove that a
certain number of test cases can be executed without bugs before the test team
accepts the delivered code (this is often called test entry criteria).

If this is not done, a large number of testers usually fail right at the same tasks (like for example no-one can install the solution), which is a large waste of resources. It is usually helpful if a smaller number of testers start with some basic test cases and, as soon as those can be successfully executed, the majority of the test team should join the phase.

The goal of the test phase is to identify bugs in the software before it is released to the end user. There are several studies that show that fixing bugs becomes more expensive the later they are found in the development cycle and especially expensive if they are found by a customer. According to Steve McConnell [8], fixing a bug that is found during the test phase could cost 10–15 times more than if that bug had been found and fixed during the implementation phase. But it can become really expensive after the product is released to the end user. Fixing defects in the field can cost 10–100 times more than fixing the same bugs in the coding phase.

I am sure everyone will agree with this statement, even without sophisticated studies, as it is simply much more complicated to debug a customer's system in production, to which the support team may not directly get access and cannot simply ask the customer to enable certain logs and traces to get a handle on the issue.

On the other hand it is also not worth while to test until there is no more defect in the product, assuming it were possible to determine when that state were achieved. The right time to hand over the solution to the customer or to release the product is usually when the defect arrival curve of a project flattens out.

There has been a lot of research on the subject of computing software reliability and there are different models available, like Exponential, Gamma, Power, or Logarithmic, to name just a few. Paul Li, Mary Shaw, Jim Herbsleb, Bonnie Ray, and P. Santhanam have compared the different models in their article [7]. For our projects we have chosen the Weibull model and are quite happy with the results (Fig. 2.10).

The challenge with most models is that someone has to estimate the overall number of defects, which is especially hard for new projects.

2.2.4.1 Development Test Ratio

Another discussion that is usually a hard debate between development and test in a traditional setup (with separate development and test teams) is the question of the right ratio of testers and developers. There is no general answer to this question. The answer is, for example, different for new projects and existing solutions that are being enhanced. In projects that are enhancing an existing solution or software product, the team also has to ensure that all the existing functionality still works. This requires additional regression test, which is not needed on an all-new project.

The test effort usually also increases the more operating systems, databases, browsers and so on a product supports. Usually the development effort for this

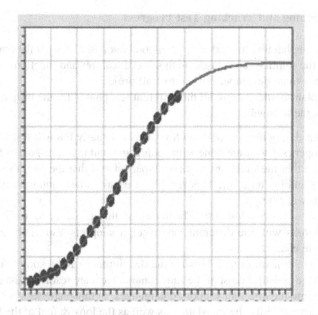

Fig. 2.10 Defect model using the Weibull distribution. As you can see in this project the defect arrival rate is following the modeled distribution

additional support (especially when Java is used) is relatively small, but the additional test effort to make sure that everything really works on, for example, a new operating system is often significant.

First, the focus needs to be on developers doing a good job in delivering high quality code to the test teams. With traditional waterfall projects there is too often the problem that the given DCUT date comes and of course the development team wants to make the date. So they rush the code into the library and build system to be able to claim that they finished on time. But now, with a bulk-load of new, potentially not well-tested code, nothing works for a few days until the team was able to stabilize the drivers. Therefore it is important to get the right mindset with the developers.

The ratio of developers to testers on the one hand really depends on the level of quality the development team delivers, and on the other hand also depends on the level of risk that the team and the company are willing to take.

With traditional projects, the developer to tester ratio is more in the area of 1:1–1:3 [13], with agile projects it is more in the area of 1:3–1:5 [16], due to the fact that more testing is required by the development teams and the traditional test teams are now more focused on regression testing and verification of additional platforms.

As it is more efficient to invest in developers delivering high quality code than in having testers find the bugs, we think that a low developer to test ratio is acceptable as long as the development teams live up to the quality.

2.2.4.2   Planning and Tracking Test Progress

The first step in planning the test is to get a good overview of what the features and functions of the solutions are. Based on this list of features and functions, an overall test plan needs to be developed for the overall project.

This test plan should contain all the different test phases for the project. Usually a project has the following:

- *Integration & function verification test:* This is usually the first test phase after the developers say they are done with coding and unit test. The goal of this phase is to ensure that the functions work as expected and that the overall solution is installable and fits together. This test phase usually also contains any required accessibility testing.
- *Globalization verification test:* The focus of this phase is to verify that the solution works with the different languages it supports, especially those with double-byte character sets.
- *Translation verification test:* Usually the development and testing is done with one language and after most of the functional defects are resolved, the product is translated into the different supported languages. In this phase testers capable of these languages verify the translation as well as the look & feel of the translated project, like, for example, does the translated message text still fit in the box the developer assigned to this message or are the last few characters missing.
- *System verification test:* Here the complete solution is tested as a whole in complex customer-like environments under load. The goal is to verify that the solution functions for several days or weeks without issues such as running out of memory.
- *Performance verification test:* The goal is to verify that the solution meets the performance requirements, especially with regard to throughput and scalability.
- *Acceptance test:* Solution development projects for a particular customer often end with an acceptance test phase during that the customer is validating if the solution meets his or her specifications and exceptions, before he or she agrees to close the project.

These different test phases are then usually planned separately, and test scenarios are developed for each of them. But it is important that there is an overarching test plan covering all the phases that ensures that all areas are covered and on the other hand also ensures that there are no duplicate test cases. There are often specialized test teams with skills in each of these test areas. The team then estimates the effort it takes to complete these scenarios. This estimate should also include the time it may take to reapply new builds and retry the test after defects were fixed. Often the test teams don't directly assign efforts in person days or person weeks, but translated them into abstract units like points.

The progress of a test is often tracked and managed using a so-called s-curve. The start of a test is usually slow until the basic problems are resolved, like, for example, installation or configuration problems that may block several test team members from carrying out their functional test cases (Fig. 2.11).

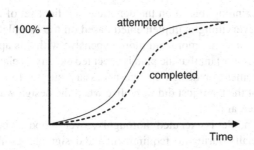

Fig. 2.11 Test s-curve

The test teams reach their highest productivity as soon as these problems are resolved and all the different test cases can be executed in parallel. Towards the end of the test, the number of blocking defects becomes smaller and there is only a small number of remaining test cases that are still failing. Therefore the s-curve is flattening out.

The test teams usually differentiate between attempted test cases, meaning the team started to test the respective test case, and completed test cases, which are those that finished without remaining bugs.

2.2.5 Support

The support phase starts as soon as the solution is handed over to the customers or made available as a purchasable product. At the same time the responsibility for supporting the product is often handed over from the development team to a dedicated support team. Depending on the size of the product, there may be a multi-tiered level of support:

- Level 1: Usually a call center with a broad set of knowledge across the complete product portfolio the company may offer. Level 1 is usually able to resolve simple problems by pointing customers to the right documentation.
- Level 2: Support specialists with deep knowledge in a product or a small number of products. They are experts in these products and can resolve the problem, unless it is a bug in the product.
- Level 3: Developers that resolve bugs in the released software by providing and testing patches.

2.2.6 Advantages and Disadvantages

There are clearly several advantages with the waterfall approach. The biggest one may be that it is really the most efficient way to carry out a project, if everything is designed at the beginning, based on the complete and final set of requirements. In the coding phase everything is implemented based on the complete and error-free design. As changes become more and more expensive with this approach the later they are done, it is important that the plan is executed exactly as planned. Every new or changed requirement or design change necessary due to, for example, design flaws or things that the team just didn't realize when the design was agreed on, is a significant incremental cost.

If the waterfall model is executed thoroughly, a very good set of documentation is produced, especially during the requirement and design phases like, for example, a detailed list of the requirements, detailed product specifications, and detailed software designs. Like all the deliverables of one phase, the design documentation needs to be reviewed as well before the project enters the implementation phase. In the waterfall approach, the test teams are usually separate from the development teams. The test team receives the documentation (and a good specification can be a clear advantage) and verifies the product against that specification. There also needs to be good product documentation right at the start of the testing phase to allow the test teams to be productive.

But it is almost impossible to carry out larger projects or projects with some level of innovation without change. It could be as simple as a client becoming more concrete with his ideas and requirements after they have seen a prototype or a first working version of the solution. In the waterfall approach, this means that the team would need to go back to the requirements and design phase to add these new requirements to the existing design and update the complete code to support the new requirements. The agile concepts accept that change will happen and therefore tackle the problem in smaller iterations and with early and continuous customer feedback.

Even today, there are certainly projects that are best done with the waterfall approach, especially small projects that are clear and mostly repeating something that was done before. If a project works perfectly fine as planned, then there is no problem. But usually this is not the case. The waterfall approach significantly limits the options, as a significant number of problems usually surfaces only very late in the project, like in the test cycle.

2.3   Project Management Triangle

The Project Management Triangle is a nice representation of the three main constraints in which a project usually finds itself (Fig. 2.12).

A change in one of the three constraints has a direct impact on the other two. If, for example, one increases the scope of the project, then this usually means an increase in cost and time. If you are limited by the time and can't finish the project on time, this could mean that you have to reduce the scope or increase the cost. The challenge for the project manager is to manage these three constraints without impacting the quality of the deliverable. The earlier he can detect a potential

Fig. 2.12 Project
management triangle

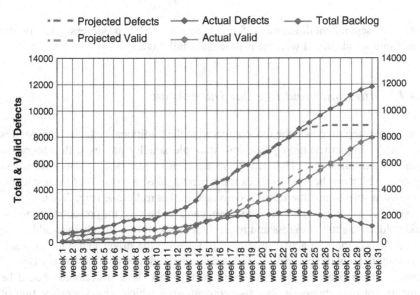

Fig. 2.13 Defect trend with a project that still has too many defects at the point in time when the software should have been ready. The projected trend is the dashed line. The planned date to have the software completely done is at the point where the dashed line starts to go flat

problem the more possibilities and time to react does he have and usually the cheaper the resolutions are.

In traditional waterfall projects, these problems usually arise the first time the project is handed over to test. Before that, the development teams may be able to cover up their problems, but as soon as a test team tries to perform some functionality it will become obvious whether a particular function works or not.

Another challenge usually arises in the test phase of a waterfall project if more problems are found than expected and the test cannot close on time.

Figure 2.13 shows an exemplary defect trend curve of a project where the time to deliver the solution to the customer has arrived, but the team still continues to find defects at the same rate as before, which is a clear indication for the product not being ready for prime time.

   In a waterfall project, the options the leadership team has in such a situation are
fairly limited. As all functions are implemented, even reducing scope is somewhat
difficult, as removing functionality usually adds extra effort. Adding more resources
to a late project can often result in a further delay if these additional resources need
to be trained by the existing team. This often leaves only two options: extending the
project until all problems are resolved and the required quality is achieved, or
scarifying the quality of the deliverable.

2.4 Modified Waterfall Models

There are several modifications of the waterfall model to address some of the
disadvantages identified with the classic waterfall model.

2.4.1 Milestones and Regular Integration

One simple step to reduce the risk compared to the traditional waterfall approach is
to split the implementation phase into multiple smaller phases with integration
points called milestones. This way, there are regular integration points with working code which can be used to check the project against the plan, and the test teams
could even start to test the features delivered in that milestone (Fig. 2.14).
   At these milestones the working system can also be used for presentations to the
stakeholders to show progress and request feedback. It may also help to add a few
freeze days before the actual milestone to stabilize the driver and only let really
important defects into the driver. These drivers could also be used for any type of
Alpha or Beta program. The important part is that these milestones should be
planned, not only date-wise, but also content-wise, which allows a calibration of
the project with regard to where it really stands. Usually, if you ask developers how
far they are with a particular task, they spend 50% of the time on the first 90% and

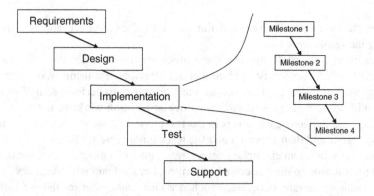

Fig. 2.14 Implementation phase with milestones

need about the same amount of time to complete the remaining 10% of the work, especially to unit test and resolve the defects found during this test.

The approach of overlapping some parts of the implementation phase with the beginning of the test phase can also be used to shorten the overall length of the project, assuming the required test resources are available.

2.4.2 Incremental Development

The incremental development approach takes the milestone approach further by not only breaking the implementation phase into smaller pieces, but by also taking the whole development process and repeating it several times on subsets of the overall project (Fig. 2.15). By doing so the whole project becomes more flexible to react to change and changing requirements. As the design for the complete solution is now done in steps and not overall at the beginning, there may be a higher risk that in a later round the designs and the code that were done before need to be changed again to support requirements that are just addressed in this round.

This approach also addresses the issue of the waterfall approach – that it takes until the test phase to see the first integrated and working code. The project is split into several increments, with each of the increments being a small waterfall project.

This way you do not have a big bang towards the end of the project, but have the project broken into smaller parts that can be managed more easily. Now it is possible to show working code much earlier to the stakeholders and customers as well as receive feedback early to have enough time to react and incorporate the feedback into the solution or product. A full development approach from requirements to test is executed in each round.

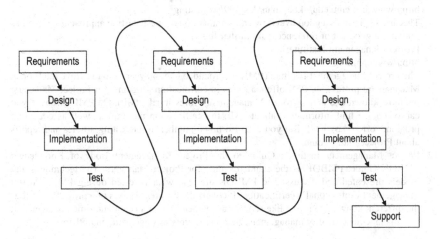

Fig. 2.15 Incremental development approach

Further Readings

1. Booch G (1993) Object-oriented analysis and design with applications, 2nd edn. Addison-Wesley, Reading, MA
2. Booch G, Rambaugh J, Jacaobson I (2005) The unified modeling language user guide. Addison-Wesley, Reading, MA
3. Centers for MEDICARE & MEDIAID Services: Selecting a Development Approach 2008
   http://www.cms.hhs.gov/SystemLifecycleFramework/Downloads/SelectingDevelopment
   Approach.pdf
   This paper provides a quick overview on some of the traditional development approaches, like waterfall, incremental, prototyping, and others and describes their strengths and weaknesses
4. Holzmann, Gerard: Economics of Software Verification
   http://spinroot.com/gerard/pdf/paste01.pdf
   This is an interesting paper on the price of defects and when the test cut off point is to release a product or solution to the customer
5. IBM Developer Works:
   http://www.ibm.com/developerworks
   This is a great site for all developers with a lot of articles, trail versions, best practices on all kind of software development topics. Under the Rational section you can for example find a lot of useful information about UML and using UML in practice as well as tools to help you with your software development project using UML
6. Jacobson I (1992) Object-oriented software engineering: a use case driven approach. Addison-Wesley, Reading, MA
7. Li P, Shaw M, Herbsleb J, Ray B. Empirical Evaluation of Defect Projection Models for Widely-deployed Production Software Systems, FSE 2004,
   http://www-2.cs.cmu.edu/~paulluo/Presentations/FSEpresent6.ppt
8. McConnell S (2004) Code complete, 2nd edn. Microsoft Press, USA
9. Object Management Group:
   http://www.omg.org
   The OMG is a non-profit organization driving standardization in the area of software. OMG specifications include CORBA but also the Unified Modeling Language (UML). On OMG's website one can for example find an introduction to UML or download the complete specification
10. Pan J. Software Testing: 1999
    http://www.ece.cmu.edu/~koopman/des_s99/sw_testing
    This article provides a good overview and summary on the basic testing approaches as well as provides a good list of references and further links
11. Project Management Institute:
    http://www.pmi.org
    The Project Management Institute (PMI) was founded over 40 years ago to evolve the Project Management profession. PMI offers a series of certification programs for Project Managers; the most known one is the Project Management Professional (PMP). On PMI's website you can of course find information about PMI, the certification program as well as education programs offered by PMI. But you can also find a number of interesting articles and reports about Project Management
12. Project Management Institute: Guide to the Project Management Body of Knowledge: PMBOK Guide. PMBOK is the standard guide for Project Managers. The techniques and models presented and discussed in PMBOK are also what is tested during PMI's Project Management Professional's certification. It covers the different phases of a project as well as the different knowledge areas, like scope management, risk management, time management, cost management, quality management, human resource management, and others

13. Rice R. The elusive tester to developer ratio,
    http://www.riceconsulting.com/articles/tester-developer-ratio.htm
14. Rumbaugh J, Blaha M, Premerlani W, Eddy F, Lorensen W (1990) Object-Oriented Modeling and Design. Prentice Hall, Englewood Cliffs, NJ
15. Software Testing Search Engine:
    http://www.qalinks.com
    This is a good starting point if you are searching for further articles and software around the topic of testing software
16. Sutherland, Jeff; Blount Jake; Viktorov, Anton: Distributed Scrum: Agile Project Management with Outsourced Development, Agile International Conference 2006,
    http://www.scrumalliance.org/resources/17

Chapter 3
Overview of Agile Software Development

Agile thinking is an attempt to simplify things by reducing complexity of planning, by focusing on customer value, and by shaping a fruitful climate of participation and collaboration. There are a vast number of methods, techniques, best practices, that claim to be "agile." In this chap. we want to give an overview of the most common ones.

The five principles of a fractal team, which we introduced in Chap. 1, apply to most of them: self-similarity, goal orientation, self organization, self improvement, and vitality are cornerstones when implementing an organization capable of executing software projects in an agile way.

The desire to establish flexible and efficient development processes which produce high quality results is not new and has not only been applied to software development:

More than two decades ago, the manufacturing industry underwent dramatic changes, when the traditional production concepts of Taylor and Ford were challenged by extremely successful Japanese enterprises such as Toyota. The Western hemisphere was puzzled at how the competition from Far East seemed to be able to produce better quality at lower cost and quickly began to outperform the rest of the world. What happened? What was the secret of the amazing efficiency and innovation?

One cornerstone of Toyota's success story was "Kaizen."

Kaizen is the Japanese term for improvement and stands for the dedication to continuous improvement throughout all aspects of life.

Masaaki Imai described "Kaizen" in a few sentences:

Do little things gradually better every day.
Set and achieve ever higher standards.
Treat everyone as customer.
Continually improve in all areas and on all levels. (Masaaki Imai [7])

This simple guideline is a driving force to adapt manufacturing processes quickly to changing customer and market requirements and incorporate a gradual never-ending improvement company-wide.

T. Stober and U. Hansmann, *Agile Software Development*,
DOI 10.1007/978-3-540-70832-2_3, © Springer-Verlag Berlin Heidelberg 2010

Toyota also focused at eliminating waste as well as out-design overburden and inconsistency from their car production (Muda). Other concepts were to stop the machine or production line as soon as a problem was detected and determine the root cause first before proceeding (Jidoka). The newly introduced Just-in-Time concept only produces what is needed at the time when it is needed. This helps to avoid unproductive queuing and buffering of resources at the interfaces between teams.

Strongly driven by an extremely successful Japanese industry, Kaizen and related approaches for organizing factories and manufacturing plants became en vogue around the globe.

In 1990, James Womack published a book with the title "The Machine That Changed the World: The Story of Lean Production – Toyota's Secret Weapon in the Global Car Wars That Is Now Revolutionizing World Industry" [16]. And the use of the term "revolutionizing" was not exaggerated: Lean production has been positioned as a synonym for the increase of productivity, flexibility, speed of turn-around, and quality by continuous improvement and continuous adaptation to a changing environment. The direct (horizontal) interaction of different organizational units as well as suppliers and customers is a key characteristic of being lean. The employee plays a significant role, as he is considered as the source of any value being created. The organizational structure and culture of lean production completely differ from the approach of traditional enterprises, which tend to be strictly hierarchical and function-oriented.

As the history of origins of lean production can be clearly found in the Japanese industry, the term "Toyota Production System" is a commonly used term as well. But "lean production" nicely points out the key message: Lean is alluding to "lightweight" and is quite the contrary to "regulated" and "bureaucratic."

James Womack came up with a set of best practices that cover technical methods as well as organizational paradigms. None of these best practices is really new. The power of lean production is their smart combination, as well as their consistent implementation within an organization. Adopting lean production implies that you are willing to change attitude and behavior within your company!

The best practices of lean production include:

- Search for the root cause of failure
- Focus on the customer value and avoid any waste
- Decentralize responsibility and accountability
- Focus on teaming and collaboration rather than splitting work
- Continuous improvement (Kaizen)
- Flexibility to react to changing customer requirements
- Standardization of processes
- Planning and anticipatory thinking
- Simple and pragmatic tools

Keep in mind that lean production was targeted for the manufacturing industry. But can it be applied to software development as well? We believe yes. The concepts of lean software development translate the best practices of lean production to the needs of IT Industry.

3.1 Lean Software Development

Based on the best practices of lean production, Mary and Tom Poppendieck [9] have shaped Lean Software Development. This approach comprises the following set of guidelines:

1. Eliminate Waste: Significant waste in software development can be introduced in multiple ways: Software engineers tend to focus too much on brilliant technologies, and often neglect the actual business value. Any development work unrelated to customer value has to be avoided. A rule-of-thumb says that 20% of your features will give 80% of a product's value. Waste also includes efforts for excessive administration and project management. For instance, you could elaborate and design more requirements than you can actually implement within the timeframe of the current project. Or you could waste time and resources by detailing future aspects, which are based on unclear assumptions and are quite likely to change anyway. If test efforts soar, you are probably testing too late. Waste is also a result of organizational boundaries, which increase costs, slow down response times and interfere with communication because of the implied hand-offs. This happens often when the development and test teams are two separate organizations and not working as one team. A typical example of waste is a feature that a developer adds to a software product only because he wants to apply a brand-new technology. The feature does not add any new attractive use cases, nor does it simplify usage. Even if the developers implemented this in their spare time on their own account, the code changes will introduce the potential of new bugs in their own area as well as in other areas and it will add the burden of future maintenance cost.
2. Focus on Learning: While planning is useful to some extent, learning is essential. Agile projects encourage prototyping as a source for feedback and improvements. A prototype is reliable input to validate designs, estimates, project progress, and customer requirements. A prototype is a good way to challenge and improve currently existing standards. Learning also includes analyzing failures, investigating their root cause, and ensuring that the same failure will not occur again. A common technique in this context is "5W": To find the root cause, you need to ask (at least) five times "Why did this happen?". This attitude will help to accept failures as a valuable pointer to improvement areas, rather than an undesirable nuisance. To give an example, a software application might crash during installation. Why did it happen?
3. Build Quality In: If your test teams find too many defects, the overall process is not working properly. A test driven development creates proper working code

from the beginning. Automated unit and acceptance tests are already part of the initial design work. Code is integrated and verified continuously by the developer who is implementing the use case. Continuous integration of code changes into a common code stream helps to avoid the painful nightmare of integrating and merging different code streams afterwards. But quality is more than just error-free code. Quality implies that the customers' expectations are met. It is necessary to understand the requirements and value proposition of the customers. And it is necessary to be flexible enough to also address very specific requirements of individual customers.

4. Defer Commitment: As no one can create the perfect plan for an entire project, the development needs to be approached in many small increments, rather than a full specification. This is difficult to achieve. It requires a break-down of dependencies between teams and components to enable them to make progress independently. The overall system architecture should support the addition of any feature at any time. Making the system change-tolerant is the key challenge that software architects are facing now. The design needs to explore options for a particular problem. Decisions that cannot be undone should be delayed until the last responsible moment, and should be made well-educated. Prototypes can be a good starting point to gather a better understanding of a subject prior to making decisions. One thing needs to be made very clear: Lean development encourages you to get started quickly and to close on the details of your project planning as you go forward. But this approach makes it even more important to start with a very clear vision and precisely defined directions that outline the scope of the work you are about to kick off.

5. Deliver Fast: Overloading teams will slow down the progress. The "Queuing Theory" applies to development teams as well, not just to mainframes: Optimizing the utilization of each individual team member creates a traffic jam that actually reduces productivity of the overall team. Instead of loading a huge pile of work onto the team, fewer things-in-process can actually increase the turnaround time. A reliable, repeatable, and sustainable pace, which limits the workload to a realistic capacity, is fully compatible with rapid delivery in high quality and at low cost. Rather than splitting work into distinct functional blocks, lean development suggests interdisciplinary teams that deliver end-to-end use cases. Each team member is in direct interaction with its customers, while the necessary interfaces and hand-over points are minimized. Standardized processes and a powerful supportive tooling infrastructure for communication and logistics are extremely helpful to accelerate turnaround. Progress in small iterations will allow making enhancements available to customers very quickly, instead of waiting for a final shipment of the next full product release (Fig. 3.1).

6. Respect People: A motivated team that is jointly working towards a common goal provides the most sustainable competitive advantage. This implies an effective leadership, which encourages pride, commitment, respect, and trust. Within such a climate, existing talent can unfold its creativity, intelligence and decisiveness. All stakeholders and participants of a successful project must act as a joint venture without unresolved conflict of interests. This implies that a

Fig. 3.1 A quick turn-around
between Design, Code,
Customer Feedback and
Refactoring & Improving
allows staying on the pulse of
the Project Sponsor

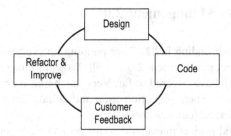

significant share of the responsibility and accountability is delegated to the teams who develop the actual customer value. Many decisions can be made best by those who have the ultimate technical expertise and who know the most recent project situation unfiltered and first hand. Encouraging self-responsibility and autonomy of teams is one key reason for the remarkable efficiency of lean enterprises.

7. Optimize the Whole: Optimizing the whole means looking at the entire value stream: What is necessary to deliver a successful and profitable product? Such a product is a unique combination of opportunity and technology and it considers primarily the needs of the customers. Outside-in design is one approach to start elaborating design with understanding the customer requirements. In order to optimize the whole, both the product itself as well as the process for developing the product, needs to be reviewed continuously. The results need to be quantified, e.g. by measuring development time or customer satisfaction. The idea of Kaizen applies to this as well: nothing and nobody is perfect! Continuous improvement in small steps, combined with iterative evolution, is the underlying driver for perfection.

Some things are really easier said than done. Eliminating waste is a good goal to aim for. But it is extremely difficult to reduce the complexity of a software stack and simplify the development process and the actual implementation. This requires a lot of experience to slice work into smaller chunks of work, and identify suitable cases that can be implemented by a team. You always have to keep the overall design flexible enough to be able to extend it at a later point in time, when the project sponsor or the team decide to change the scope. While it would be helpful to remove existing features which no customer is using from a product, it is difficult to find out who is using what. Often, we will know that customers at least tried to use a certain feature if they call in to report a problem with it. But we rarely hear about features that customers are using without problems.

Another difficulty is the availability of suitable tools that support the necessary decentralization of typical project management and planning. We also need powerful tools to support the collaboration and communication within and across teams. Promising solutions can be found around the Web 2.0 initiatives.

3.2   Project Management 2.0

Since buzzwords ending in "2.0" are popping up everywhere, it is no surprise that there is a project management 2.0 as well. Project Management 2.0 covers many of the thoughts we have covered so far. Very much like Web 2.0, it puts a strong focus on social aspects, such as participation and collaboration, rather than centralized control and hierarchical structures.

In traditional project management, the project manager gathers the status information, processes it, and reports the results to upper management. In large development projects, the amount of information can easily begin to overwhelm the project manager. Project management 2.0 delegates a significant amount of project management responsibility to the teams. Teams can manage their own realms pretty much self-contained by creating a collaborative space, in which every team member is able to contribute both project deliverables as well as project management-related information. Each team member will know about any pressing problems first hand. A significant benefit of this approach is that the expertise and experience of each individual can be applied best to benefit the overall project. The "wisdom of the crowds" is the key source of knowledge, which makes all these "2.0" initiatives so powerful.

A nice example is the planning poker, used to create effort estimates: Each participant in the poker game "votes" by laying cards representing their sizing estimate face down on the table. This process is repeated until a consensus is reached.

But project management 2.0 it more than just tapping collective intelligence: Project progress is made clearly visible to everyone in the team and results in transparency. With short iterations and a working solution at the end of each iteration, progress is also made extremely visible to the project sponsor as well as the other teams working on the same project. The project manager can focus on leading the teams towards a visionary goal rather than focusing on counting people days and fitting work hours into a project schedule.

Such a collaborative environment can be implemented utilizing easy-to-use and strongly decentralized second generation web-based tools, such as a Wiki, social networking platforms, and other web 2.0 related technologies. But there is also a growing number of specific project management tools, which are tailored towards project management 2.0.

3.3   Agile Manifesto

In 2001, several agile thought-leaders agreed on what they called the Agile Manifesto:

> We are uncovering better ways of developing software by doing it and helping others do it. Through this work we have come to value:
> *Individuals and interactions* over processes and tools
> *Working software* over comprehensive documentation

Customer collaboration over contract negotiation
Responding to change over following a plan
That is, while there is value in the items on the right, we value the items on the left more. [4]

These principles govern all the techniques and rules in the different agile methods, which all strive to make software development more flexible and overall more successful.

3.4 Scrum

Scrum was introduced by Takeuchi, DeGrace, Schwaber, and others in the late 1990s [11].

The concept behind scrum is drastic simplification of project management. Scrum is a rough outline of a process, based on iterative development. It comprises of three roles, three documents, and three meetings.

The three *roles* are product owner, team, and scrum master:

The product owner represents the stakeholders, such as customers, marketing, etc. The product owner has to ensure that he or she is representing the interests of all stakeholders. He is also providing the requirements, funds the project as well as signs off on any deliverable. The development team is just called team. It is responsible for developing and testing the project deliverable. And finally there is the scrum master. He is responsible for the scrum process and for adjusting scrum to best fit to the project and the organization, as well as to ensure that any issue or problem gets resolved and that the team can be as effective as possible.

The most important role of the scrum master is to ensure that the team is really delivering high quality and is not cutting quality to get the job done. The scrum master has to check the team, verify that the test cases and code reviews are done, and must have the standing to stand in front of the team and the project sponsor and tell them if the team is scarifying quality in favor of features. There may be a short-term success with cutting quality in a release and getting it out of the door on time, but by doing so you take out a credit on the next release of the project. In the next release the velocity will be worse because of all the problems that were left in the solution. If the team does not take the time to fix them now, then the problem will become even bigger – but acknowledging that the last release was not of high enough quality will not be popular at all. But if this is not done, the productivity will go down from release to release and the problem will just become bigger with every release. Especially in a larger project where other teams depend on the quality of the code that is delivered, like test teams doing system verification test type of use cases, this has a significant downstream impact also on the current release. Each problem that, for example, a test team finds again in the code is wasted time that could have been saved by immediately fixing the problem (or not putting the bug in the code at all).

Let's take the example in Fig. 3.2. Let's assume the team is able to get ten "things" done in an iteration. Now they are told they need to deliver more

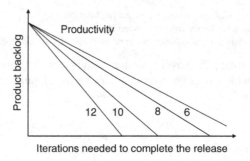

Fig. 3.2 Productivity decreasing from release to release because of bad quality delivered in the release before

functionality, and to get this done for a given release date, they need to get 12 "things" done per iteration. They can try to cover some of the extra work using overtime, but in the end they will cut quality to get the job done. In that case, they successfully deliver the release on time with the requested features and are asked to do the next release. As they did such a great job before, the customer or the product management team is now asking them to get even more done in a shorter time. But, due to the bugs that are remaining in the code, the team can now only get eight "things" done per iteration. This way, the productivity would decrease from release to release, unless someone stands up and stops the death spiral. Later in this chap. we introduce velocity as a way to measure the productivity.

It is the role of the scrum master to ensure that there is high quality from the beginning and that no-one cuts short on quality. This is not an easy role at all, but a very important one for the success of the team and the product.

There are three important *documents*: product backlog, the sprint backlog, and the sprint results:

As with traditional projects, all the requirements are gathered and prioritized in one list, which is called the *product backlog*. The development is done in small iterations, which are called *sprints* in scrum terms. Each sprint is a small and manageable iteration. It contains design, development, testing, and documentation. The duration of a sprint is usually about 2 weeks. The goal is to produce a tested and stable deliverable which can be handed over to the customer and put into production by him. At the beginning of a sprint, the team picks the most important use cases that can be delivered in that current iteration. The list of those use cases is referred to as *sprint backlog*. The sprint backlog is negotiated between the product owner, the scrum master, and the team itself. The use cases that are completed during that sprint are documented as *sprint results*.

The three *meetings* are the sprint planning meeting, the daily scrum meeting, and the sprint review:

Each new sprint starts with a planning meeting. The team, the product owner, and the scrum master decide on the sprint backlog, based the requirements prioritized by the product owner and the estimates the team has done and therefore on what the team can commit to deliver in one sprint. This meeting consists of two parts: In the morning, the product owner presents the most important requirements from the product backlog to the team. At the end of the morning, the team selects those items that are going to be delivered in that sprint. In the afternoon the team plans the sprint in more detail.

Throughout the project, the scrum master moderates a short 15 min project meeting (called the daily scrum meeting) every day to exchange information what was achieved in the last 24 h and to discuss problems that need to be addressed. It will be discussed ...

- ... what each team member achieved during the last days,
- ... what issues occurred and may lead to reconsidering the iteration plan or design,
- ... what each team member will be working on during the next days and
- ... what issues and road blocks need to be addressed.

The major goal of these daily meetings is to determine the progress that was made and identify any issues that need attention or require adjustments of the sprint backlog. Based on the progress, additional items can be added from the overall product backlog to the sprint backlog. Or – in case of delay- items of the sprint backlog can be moved back into the product backlog to be addressed in a later iteration.

At the end of each sprint scrum master, team, and product owner (along with the other stakeholders) meet again to review the results achieved during the iteration. This meeting is referred to as sprint review. We found it very useful to have this meeting go along with some demos. Especially in a large dispersed team it is important that all team members know what is going on and where the team stands compare to the backlog for the current iteration (Fig. 3.3).

The horizon for detailed planning is short and only covers a single iteration. Instead of building a large plan which includes all the tasks, estimates, and

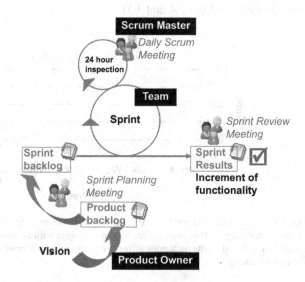

Fig. 3.3 Scrum (*Source: Joseph Pelrine, MetaProg*)

resources for the entire project, only very rough high-level estimates are done at the beginning of the project.

More detailed estimates and work assignments are added at the beginning of each sprint only for those items that the team has picked to be delivered in that particular sprint.

This is a major difference to a traditional waterfall project, which elaborates plans and milestones to a much larger extend. This aspect of the waterfall approach makes changes of the finalized plan quite inflexible and expensive, especially late in the cycle. In a scrum-based approach, the project management as well as the product owner and the other stakeholders accept this change as a fundamental planning constraint and do not waste effort in planning beyond the horizon of the current sprint, since there are too many uncertainties. A long-term plan would be subject to changes anyway. And most important: Such a well-defined plan for the future can only give a misleading and false confidence that no-one should trust.

Agile approaches like scrum are addressing this problem by breaking a larger project into iterations and really only worry about the content of the current iteration.

Scrum gives the opportunity to make fact-based decisions for the immediate needs. And it allows to even change the fundamental direction of the overarching direction at the beginning of every new sprint if this becomes necessary, to reflect changed requirements and add actions to address unforeseen issues.

Scrum recommends tracking the progress of the development activities and to measure progress towards achieving the given goals using burndown charts. The amount of work a team has left to complete within the current iteration is listed on the vertical axis, whereas the time is shown on the horizontal axis. The remaining work can be shown in units of hours, person days, or number of use cases. During each of the daily scrum meetings a new value is added. The resulting graph shows the team's progress towards iteration end. The following figure show quite different burndown charts (Figs. 3.4 and 3.5).

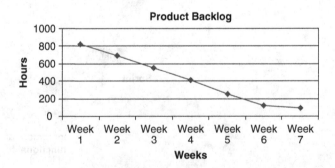

Fig. 3.4 This burndown chart shows an ideal world: The team is making constant progress towards the end of the project. This chart covers the entire project with all intended iterations. The time to completion is based on the remaining effort to complete the final product with its anticipated scope ("product backlog")

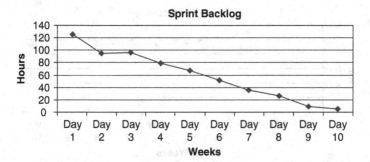

Fig. 3.5 This chart shows the remaining effort for completing the iteration with its intended content ("sprint backlog"). In this example, the team has made no progress between Day 2 and Day 3. The reason could be that the team members had to help out another team and could not focus on their own tasks. It could also be that it took longer to complete the first use cases of the iteration and the start of the next one was delayed. Starting with Day 3 the team could catch up again. Either because they could complete the remaining work faster than expected or because the sprint backlog was reduced by deferring some use cases

In a smaller project with only one team and not more than eight to ten team members a regular burndown chart usually gives you enough detail to judge the status of the project, as you have made sure that tasks and user stories are small in size (a matter of days or a week maximum) and that the team is continuously completing tasks and user stories before starting new ones. What you want to prevent is that in, let's say, a 6 week iteration the team is coding the first 4 or 5 weeks and then tries to rush through the test in the remaining 1–2 weeks. The regular burndown chart will still show you nice progress, but at the end of the iteration you may end up with nothing, as it may happen that the team was not able to complete all tests and fix all necessary defects.

Especially for larger projects with several different teams we recommend a separate burndown chart for tasks & user stories, especially for those that may be distributed around the world and just don't allow you to be deeply involved in the details of every team's progress. A task or a user story is only counted as complete if it is completely done including automated test cases, all tests executed, all documentation done, and all remaining defects fixed. You can see an example of such a burndown chart in Figs. 3.6 and 3.7.

You clearly see that the team may have done a lot of work and the regular burndown chart may look much better, but the team back-loaded the iterations too much and wasn't continuously completing tasks and user stories. The team may have coded a lot at the beginning of the iteration but ended up not being able to finish the user stories until the end of the iteration. A regular burndown chart could also give you indications, but usually way too late to turn the ship around.

Agile software development has several challenges, especially for the project managers: Stakeholders typically want to know when the project is completely done and what exactly will be delivered. In many situations the customer will only accept a fixed-price contract which requires the software development company to

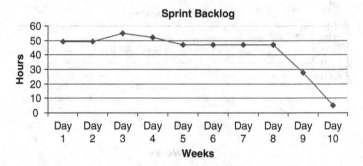

Fig. 3.6 This chart exposes some significant problems. There is no progress measured until the 9th of May. In contrary, some unexpected additional work increased the remaining efforts on the 2nd of May. The steep reduction of the backlog towards the end of the iteration is probably supported by removing items from the sprint backlog

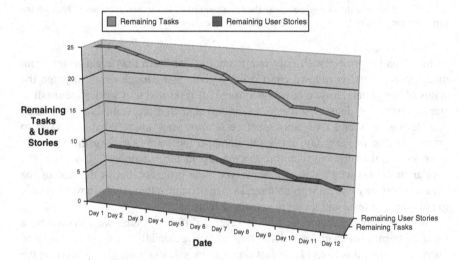

Fig. 3.7 Remaining tasks and user stories

have a very good handle on the total cost of the project; even if it is not exactly planned until the very end.

Does agile software development allow for a reliable management of overall cost and content at all?

First of all, it is important to emphasize the limitations of the pretended precision achieved with the classical waterfall model: The elaborated project plan of a complex IT project can only make one believe in exact predictions if one expects that no surprises will happen between now and the project delivery. Trusting a plan beyond the horizon of the next few weeks leads to dangerous blindness. In the final analysis, the predicted cost is still significantly different from the estimate, even if a project is planned to the very last single task. Software development is far too complex. Things will change during the course of the project. Customers can change their minds about what they want or need and usually the technology is providing more challenges than expected.

In all software development projects, all involved parties need to accept that unpredicted challenges will occur and the final deliverable cannot be fully understood at the beginning of the project. Therefore the focus of scrum is not on perfecting the precision of planning, but rather maximizing the ability to adjust planning and respond to change.

The focus of the involved parties is on negotiating only the high-level requirements and major usage scenarios upfront. While a number of key use cases will be committed to the stakeholders, there needs to be some flexibility to allow new or additional use cases to be added and others to be removed. At the end there always needs to be a compromise between the development team and the stakeholders. The development team commits to certain high-level requirements within a given timeframe. But the stakeholders, especially the product owners, also need to provide the development team with some level of freedom to refine use cases depending on the actual progress.

In scrum, each sprint is 2–4 weeks long. Our experience is that the length of the sprint or iteration also needs to be reviewed at the beginning of a project and possibly adjusted, for example, based on how long it takes to build the whole product. A complex product that needs several hours to get built may require a sprint of, for example, 5 or 6 weeks, just due to the slower turn-around time. Here you also see that a good infrastructure that can turn fixes around quickly and provide them to test in a matter of minutes or hours is an important component to run an agile project or to make a project more agile. Needless to say it is important that each sprint has a predefined fixed length and everything that was not done in that sprint automatically moves out of this sprint. There should not be any items that are "almost" done and need to be finished in the next iteration, unless this "almost" done part is a scenario on its own, production-ready and can bring value to the customer.

3.5 Test Driven Development

Automating all the test cases is a key factor to success with improved productivity in a software development project. Rerunning manual test cases several times to ensure that nothing broke (for example at the end of each iteration) is simply too labor-intensive and therefore too costly.

Test driven development is a development method to ensure that the developer also focuses on the development of the test cases. The developer starts with the design of the additional functionality to be implemented. As soon as the design is done, the test case is written – before the functionality itself is implemented. While implementing the test case the developer also validates the design with respect to what the function needs to provide, and how it needs to behave. A test case should

Fig. 3.8 Test driven
development cycle

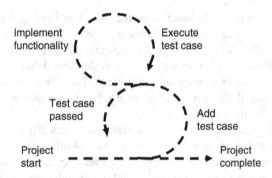

only be a relatively small increment that can be developed and tested in a matter of hours. After the code has successfully passed the test case, the cycle starts again with an additional or enhanced test case, followed by the implementation of the function, until the test case passes.

The developer starts to implement the code as soon as the test case for a use case or scenario is finished. He now has the benefit that he can directly run the automated test case against the new code to ensure that it passes the test or identifies areas that still need to improve (Fig. 3.8).

This should also ensure that there is good test coverage for all the code areas developed and that only the code is written that is needed to pass these test cases (all additional code would be considered "waste" in lean software development terms). If the functionality is implemented first and the test cases are done later, there is always the risk that automated test cases are only provided for a subset of the functionality, and that at a certain point in time, the developer has no more time left or just decides that the test cases are done.

With the test driven development approach it is especially important that the test suite can be executed within a few minutes to provide instant feedback. If the test suite for the complete project becomes too large, it may need to be broken down into subsets that each developer can execute within minutes, and the complete test suite should then be executed in a separate environment several times a day to ensure that no other part of the product broke because of a change somewhere else.

3.6   Extreme Programming

The first book on extreme programming (XP) is titled Extreme Programming Explained and was published by Kent Beck in 1999. This book is based on the work Kent Beck, Ron Jeffries, and others did together while developing the Chrysler Comprehensive Compensation System between 1996 and 2000 [6].

The 12 concepts behind extreme programming are not really new, but have been brought to the next (extreme) level:

- Test driven development
- Pair programming
- Refactoring
- Simplicity
- Planning game
- Small releases
- Continuous integration
- Continuous testing
- Collective code ownership
- Sustainable pace
- Coding standards
- On-site customer
- Story cards
- Prototype UI and UI navigation
- Stand-up meeting
- Iteration completeness/timeboxing

The terms may vary slightly depending on whose book or article about extreme programming you may be reading, but the ideas and concepts are the same.

We covered test driven development in the previous chap., so let's go directly to *pair programming*. With pair programming, two developers are working as a team at one computer on the same task. This actually saves cost and time, as two heads write better code than just one, and four eyes see bugs earlier than two. Studies have shown that the same amount of code is produced, but the quality is higher. Pair programming usually requires some practice to become really effective and may seam a little bit strange to the developers at first.

Refactoring is improving the code without changing the outside behavior of the component. Refactoring should be done in small steps to, for example, remove code not used (anymore), simplify the flow, or make the code more readable. After each step the automated test suite should be executed to ensure that the component still functions as needed. The small steps should ensure that mistakes are identified early and can quickly be fixed.

Refactoring should be a standing item on the ToDo list, especially with a product or project that has been running for several years and has delivered several releases. Over time, new features and ideas are added which may, in the end, not fit very well anymore with the original design. In addition, features that may have been requested and delivered in an earlier release may not be needed anymore and should be removed. Code that doesn't exist cannot fail or produce other problems. Static code analysis tools and refactoring tools in several Integrated Development Environments (like Microsoft Visual Studio or Rational Application Developer) will also help to identify areas that would benefit from refactoring or will suggest automatic refactoring. It is usually not easy to build a business case for refactoring, but it pays off. Fewer tests are needed, the compiled code is smaller and therefore faster to install, it uses less memory, and is easier to maintain and service by a support team.

The idea behind *simplicity* is to ensure that the development team focuses on the functionality that is needed and that no unnecessary code is added (that may be needed in a week or a month or potentially never, as requirements may change over time). This is also similar to the lean development goal of not producing any waste.

Planning game – At the start of the project, the customer describes the required features and their respective priorities, the team identifies the tasks needed and, based on this, does a rough estimate for each feature. Based on this information, the customer may adjust the priority. All the estimates are rough, preliminary, and may only give an indication of when the project will have completely implemented the requested features. XP projects are usually done in 2-week iterations, delivering customer-ready code at the end of each iteration. The team meets with the customer on the first day of an iteration, identifies the features that will be implemented in the next iteration, and presents the results to the customer at the end of the iteration.

Small release – Each iteration should deliver a customer-ready and useful system. In the extreme, the customer should be able to stop the project after each iteration and still have a working system, or they should be able to completely change the priority order of the features from the next iteration on. It is also important to note that all use cases targeted for an iteration need to be 100% completed in that iteration. An additional benefit is that you can quickly show something, experience immediate success, and can better estimate where the project is time-wise. In traditional projects, teams often work for months on their isolated parts before everything comes together. This can lead to surprises late in the project – something that could be prevented with small releases that force continuous integration. A complete product build should happen at least every day, and should be followed by a complete run through the automated test suite.

Continuous testing ensures that the code is really doing what it is supposed to do and confirms that none of the existing functionality broke with the latest changes. An extensive set of automated test cases is needed that the developer can run during unit testing, and that can also be executed against the (at least daily) product builds. As soon as the developer changes code, the test cases must be adjusted or new ones written to cover the additional functionality. This should be easy to achieve for internal modules with no user interfaces. There are also several tools on the market to test user interfaces (UI). However, the maintenance for these test cases is usually higher than a rewrite, because they often require adjustments as soon as the look & feel of the UI changes as slightly as a button being moved just a bit.

Ownership means that all developers know the code and should be able to work on all parts of the project. The benefits are that there is better code quality, as several people look at the code, the knowledge about the code is better distributed among the team, and therefore more people can help out in "emergency" situations. There is of course a limit to collective code ownership – as soon as the project team becomes larger than 10–12 developers, a subset of the team should collectively own a certain part of the code.

Teams that continuously work at an unhealthy level of overtime are not productive in the long run. Overtime is a good method to finally get some work done, but this period of the project needs to be limited to a few weeks. Research has shown that it is more effective to work at a *sustainable pace* that the team can keep for the whole project.

Coding guidelines or standards are important for every project to ensure maintainability and code quality. These guidelines can base on examples available on the Internet or in books, but need to be extended with project-specific conventions and naming guidelines.

The project team always needs to keep in close contact with the customer to ensure that the project is meeting the customer's expectations at any given time (called *on-site customer* in XP terms). This is also necessary to ensure that any change in the customer's requirements is directly reflected in the project. Developers usually only have a limited domain knowledge of what the product will be used for, therefore the customer representative should be someone who is good in translating the requirements into a form that developers can work with.

In our experience with extreme programming, some of the ideas work better in smaller teams than in larger teams.

The following XP ideas also work well in larger teams:

- Test driven development
- Pair programming
- Refactoring
- Continuous integration
- Continuous testing
- Sustainable pace
- Coding standards
- On-site customer

The following ideas may provide challenges when applied to larger-scale projects:

- Small releases
- Simplicity
- Collective code ownership
- Planning game

Small releases: Usually XP iterations have a duration of about 2 weeks. This works fine in small, ideally new projects that do not take too long going through the build. In our project, the build currently roughly takes a total of 16 h. It consists of several separate pre-requisite builds, which are then merged into the product build, which in turn produces the installation CDs for ten different platforms. Obviously, having several prerequisite builds is not ideal, and putting them all in one code library with one build would improve the situation, but this would take time and effort, since the different builds comes from different heritages. Then again, in larger projects with a better build structure, doing a complete product build in less than an hour is seldom possible. This just does not provide enough possibilities for quick turn-around of feature and fixes and therefore does not allow for having 2-week iterations. We have therefore chosen iterations that are between 4 and

6 weeks long, starting with 6-week iterations, and doing shorter ones towards the targeted last iteration.

It would of course be ideal if everyone knew every piece of code in a larger product, but it is just not realistic in a large project with teams located around the world. Then again, you may not want everyone changing code in every place of the product (as the product is just too large for one single person to be involved deeply enough in every single piece of the code). It is good to have team ownership for certain areas of the code. As an example, take the layer accessing the database (and we support a large number of different databases). You do not want to have every team write their own database layer, on the other hand you do not want everyone to change it without someone ensuring that the changes are compatible and still ensure optimal database performance. For these reasons, the database layer in our case is owned by one team during the entire whole project.

Simplicity is a very good goal to aim for, but the problem in larger projects with several different teams is often the dependencies between the teams. Of course this is not ideal, but as mentioned above, collective code ownership is usually simply not feasible in a large worldwide project. This can mean that the database team mentioned above may have to do changes in the database layer that are prerequisite for another team. These changes are often done in the iteration before the other team needs them. In lean terms, this is obviously "waste" and against the idea of simplicity. However, it was the best compromise we were able to come up with.

It would obviously be ideal if every team member could participate in the planning game with the customer or project sponsor, but these meetings tend to become ineffective if held with more than eight to ten participants. It is therefore considered a good compromise to have the team leads or architects do the planning sessions for the product, as well as the detailed iteration planning sessions with their teams. To ensure that nothing is lost in the communication from product owner to team lead/architect to the team, you may want to have the product sponsor, or at least someone with domain knowledge, participate in every team planning session.

3.7   Rational Unified Process

The Rational Unified Process (RUP) was developed by Rational Software under the lead of Ivar Jacobson, Grady Booth, and James Rumbaugh in the mid 1990s.

3.7.1   Best Practices

It defines six best practices that the authors of RUP observed as commonly used in the industry. These best practices are:

- *Develop software iteratively:* As with several of the other approaches, RUP suggests to do the development in small chunks aka iterations with a running solution or product at the end of each iteration. This allows for minimized risk and a closer feedback loop with the stakeholders and sponsors, as they can actually see working code every 2 weeks.

- *Manage requirements*: Capture requirements as use cases and scenarios, organize and track them, and document decisions and the reasoning for them.
- *Component-based architectures:* A component is a software unit that can perform a certain number of well-defined operations. RUP recommends developing a flexible architecture that can accommodate change and foster reuse of these components in different parts of the projects or, even better, across several projects.
- *Visually model software:* The recommendation is to visually model the software to allow for a better discussion of the design and illustrate how the different components of the solution will work together.
- *Verify software quality*: The quality of a solution needs to be verified to ensure that it meets the functional as well as the performance and load requirements. The Rational Unified Process supports the planning, design, implementation, execution, and verification of the test.
- *Control changes to software:* Changes to the code need to be controlled for several reasons. The most important one is that changes need to be tractable, especially if a problem is suddenly introduced and the team needs to find out what changed to identify the root course of that new problem. As soon as the solution is rolled out to the customer, the team needs to be able to identify the exact level of the source code that was released. RUP describes how to control, track, and monitor changes for successful development in iterations.

3.7.2 The Phases

Each development cycle in the Rational Unified Process has the following four phases:

- *Inception phase:* In this phase, the requirements and goals are defined and a business plan established. This includes creating all the use cases that define all the interactions with the system. Another deliverable is a business plan that also contains the high-level project plan, estimates and resource needs, major milestones, and a risk assessment. The lifecycle objectives milestone marks the end of the inception phase.
- *Elaboration phase:* The architecture for the project is put in place, a detailed project plan is created, and the architecture decisions for the overall system are made. The lifecycle architecture milestone with a detailed review of the system objectives, designs, and major risks concludes this phase.
- *Construction phase:* In this phase, the solution is implemented and tested. The focus is on managing the implementation, tracking development and test progress, tracking the schedule and the quality of the delivered code. The initial operational capability milestone concludes this phase. At this point in time, the team needs to decide if the features and the quality of the solution meets the requirements and if it is ready for a first user test, often called Beta test.

- *Transition phase:* During this phase the first users are working with the newly delivered solution and are providing feedback, problem reports, and new requirements to the developers. This could, for example, also be a phased roll-out where the basic functionality is working and provided to the end users. Later in the phase, more and more advanced features are added, and more end users are working with the new solution. Resolutions to larger problems are phased in quickly, based on their severity. Feedback and new requirements from the initial users may be incorporated before general availability of the solution, especially if they concern usability problems or slow performance of the system, others may be postponed to a later release. The transition phase ends with the product release milestone, at which the current project ends. Another release of the solution may follow, and the inception phase of that later release may have already started in parallel to the current transition phase.

Iterations are an integral part of the Rational Unified Process, with each of the four phases being split into several iterations, as advocated previously – and because you want to have a working solution every few weeks to be able to react quickly and flexibly to changing requirements.

3.7.3   The Process

The following four elements are the base of the process:

- *Worker:* Workers are the roles that an individual has in the project. This may or may not be equivalent to a physical person, as someone may have two roles in a project, such as designer for a specific function and at the same time tester for another function.
- *Artifact:* Artifacts are the models, designs, pieces of code, executables and other things that are produced in this project. An artifact can be the input for an activity or the output of it.
- *Activity:* An activity is a task that a worker performs as part of this project. Activities are usually in the size of a few hours to a small number of days.
- *Workflow:* A workflow is a series of activities during which workers produce a significant artifact as part of this project. It is more than just a number of activities; a workflow structures the activities to produce a valuable artifact. The Rational Unified Process defines a number of core workflows that we will discuss in the following paragraphs.

3.7.3.1   Core Workflows

The Rational Unified Process defines a total of nine workflows to structure the project work. Six of these workflows are considered "engineering" workflows, three are "supporting" workflows.

The six engineering workflows are:

- Business modeling workflow
- Requirements workflow
- Analysis & design workflow
- Implementation workflow
- Test workflow
- Deployment workflow

These are the three supporting workflows:

- Project management workflow
- Configuration and change management workflow
- Environment workflow

The importance of each workflow at a certain point in time changes during the course of the project, but most of the workflows are executed in parallel and not in sequence, as is the case in projects using the waterfall model.

Engineering Workflows:

- *Business modeling workflow:* The output of this workflow are the business use cases which describe the business processes that the solution needs to support. The workflow also helps with the communication between the business owner and the software development team to ensure a common understanding of the business processes and scenarios that need to be supported.
- *Requirements workflow:* The requirements are put in writing and agreed upon in the form of use cases that describe the interaction between the users and the system as well as the interaction of the system with other systems that generate input or process any output generated by this solution.
- *Analysis & design workflow:* During this workflow the design model and the analysis model are generated, which together describe how the solution will be implemented to meet the outlined requirements.
- *Implementation workflow:* The design is translated into working code. This also includes unit tests and the integration of these units into a working system. Keep in mind that RUP promotes an iterative approach were design, implementation, and test is done in iterations throughout the project and continuously produces a working system.
- *Test workflow:* This phase covers the verification of the implemented system against the requirements and use cases. The test needs to verify that the functionality is working as required, that the system performs reliably over a larger timeframe and that the required performance characteristics are met.
- *Deployment workflow:* This contains the packaging of the software and making it available to the customer through the various distribution channels that the product may be offered through, as well as putting in place procedures for customer support and problem fixing in the field. In a customer solution project this may be covered by two phases: The first one ends with the customer accepting the deliverable. Then the maintenance phase starts, which may or

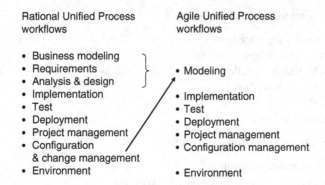

Fig. 3.9 Processes of the Rational unified process compared to the processes in the Agile Unified Process

may not be part of the project. It many cases, the complete project including source code is handed over to the customer and the customer provides the maintenance and the further enhancements with their own staff.

Supporting workflows:

- *Project management workflow:* This workflow focuses on the project management activities required to manage the project in an iterative way and ensure a timely delivery.
- *Configuration & change management workflow:* It is important to be able to identify "What has changed?" at any given point in the project to identify actions that may have contributed to regressions. Many of those changes need to be managed with the appropriate tooling that handles simultaneous updates to the same units, allows Multiple concurrent development streams, and provides the ability to exactly identify which level of code is in which build level.
- *Environment workflow:* The focus here is on providing an efficient development environment with tools and processes that support the project. This contains but is not limited to build environments, source code control systems, defect management systems, sandbox environments for the developers, software modeling tools, functional and performance test tools, and so on.

3.8 Agile Unified Process

The Agile Unified Process (AUP) is a derivative of the Rational Unified Process. AUP was developed by Scott Ambler between 2002 and 2006 and combines some of the core workflows from RUP. AUP combines the business modeling workflow, the requirements workflow, and the analysis & design workflow into one single model workflow, which also contains the change management part of RUP's configuration & change management workflow, which in AUP is now the configuration workflow. The reason behind moving the change management workflow into the modeling workflow is that in agile projects the requirements are reviewed at the beginning of every iteration, and new requirements are added or existing ones modified. Therefore it makes sense to consider change management as part of the requirements discussion in the modeling workflow (Fig. 3.9).

The phases are identical to those of the Rational Unified Process.

The principles in the Agile Unified Process are very similar to the principles in the other agile approaches, like agility, simplicity, do the highest prioritized requirements first, and trust the team.

3.9 Agile Model Driven Development

Scott Ambler [12] has taken the concepts from the model driven development (MDD) and brought them to the agile level. Models are also an important piece of the Rational Unified Process, where models are developed in an iterative fashion, compared to MDD, which is usually used more in traditional development approaches.

The agile model driven development (AMDD) concept takes the approach we have described under test driven development and applies it to models: The team starts with a very limited model which is good enough to allow them to do rough estimates. In the iterations the developer then implements the model in code, enhances the model, and implements the next level of code and so on.

The Agile Model Driven Development defines a number of best practices you will now already be familiar with from the agile approaches we presented previously:

- *Active stakeholder participation:* The customer, stake holder, or product owner needs to be heavily involved in the project to be able to provide the information and details that the team needs in a timely manner.
- *Architecture envisioning:* When the team starts, it needs to first sit together and sketch out and agree on a high-level architecture on which it then bases their other estimates.
- *Document late:* Documentation is important and needs to get done, but should be done at a point where what is documented is solid enough to ensure that as little rework as possible is done in the documentation. In our experience, documents need to be an integral part of every iteration. A user story is only complete when the documentation is done, otherwise it may never happen.
- *Executable specifications:* Define specifications as user stories that allow for easy translation into test cases.
- *Iteration modeling:* Enhance your model at the start of every iteration of part of your planning activities.
- *Just barely good enough (JBGE) artifacts:* A model should be exactly at the level that it needs to be at for the next iteration. The model may have to change in the next iteration, and models that go beyond the current iteration may need to be changed.

- *Model a bit ahead:* At the same time you may want to look down the list of priorities to see what comes next and ensure that your model allows for easy addition of these requirements.
- *Model storming:* Before you implement the next piece of code in an iteration, take a look at the model as well, and enhance it as you go.
- *Multiple models:* Pick the type of models that best fit your project.
- *Prioritized requirements:* Build a prioritized requirements list or what is also called a product backlog and start with the implementation of the most important requirements first.
- *Requirements envisioning:* At the beginning of the project, the team should meet with the product owner and build the product backlog with the currently known requirements and put them in the order of priority.
- *Single source information:* Use one repository to store the information related to the project to allow everyone easy access.
- *Test driven design:* Use test driven design; develop the test case first, before you start coding the functionality.

Further Readings

1. AgileSoftwareDevelopment.com
   http://agilesoftwaredevelopment.com/
   This is a community website where you can post ideas, comments, or articles around Agile Software Development. Some of them highlight interesting aspects and good evolutions
2. Agile Unified Process
   http://www.ambysoft.com/unifiedprocess/agileUP.html
   This is Scott Ambler's homepage. He derived the Agile Unified Process (AUP) from the Rational Unified Process and his website contains extensive of information and details on the Agile Unified Process
3. Ambler S (2002) Agile modeling – effective practices for extreme programming and the unified process. Wiley, New York
4. Beck K, Beedle M, van Bennekum A, Cockburn A, Cunningham W, Fowler M, Grenning J, Highsmith J, Hunt A, Jeffries R, Kern J, Marick B, Martin RC, Mellor S, Schwaber K, Sutherland J, Thomas D. The Agile Manifesto: 2001,
   http://agilemanifesto.org/
5. Chrysler Comprehensive Compensation System:
   http://calla.ics.uci.edu/histories/ccc/
   Here you can find a good summary on the Chrysler Comprehensive Compensation System, including some numbers about its size
6. Herela H. Case study: The chrysler comprehensive compensation system 2005
   http://calla.ics.uci.edu/histories/ccc/
7. Imai M (1986) Kaizen: The Key to Japan's Competitive Success. Random House, New York, NY
8. Cohn M. Mountain goat software:
   http://www.mountaingoatsoftware.com/
   Mike Cohn is an active promoter of scrum and has published a number of books on the subject. This website contains interesting papers and presentations on Agile Software Development. You can also order planning poker cards or can play planning poker online

9. Poppendieck, Mary, and Tom:
   http://www.poppendieck.com/
   Lean Development Software - An Agile Toolkit for Software Development Managers. Addison-Wesley Longman 2003 Implementing Lean Software Development: From Concept to Cash. Addison-Wesley Longman 2006 Mary & Tom Poppendieck's homepage and books are a good place to start if you want to learn more about Lean Development. In addition, a collection of their essays are provided that you can read online

10. Rational Unified Process: Best Practices for Software Development Teams:
    http://www.ibm.com/developerworks/rational/library/content/03July/1000/1251/1251_best-practices_TP026B.pdf
    This is a very good paper summarizing The Rational Unified Process (RUP)

11. Schwaber K, Beedle M (2001) Agile software development with scrum. Prentice Hall, Englewood Cliffs, NJ

12. Scott Ambler: Agile Modeling Homepage:
    http://www.agilemodeling.com/

13. Scrum Alliance:
    http://www.scrumalliance.org/
    The Scrum alliance is trying to build a community around scrum as well as a good resource for articles, list of courses, and news on scrum

14. Scrum et al. – by Ken Schwaber:
    http://video.google.co.uk/videoplay?docid=-72301 44396191025011&ei=T811SYrJOonojg-Lngsm5BQ&q=scrum
    This is a good video of a session Ken Schwaber did for Google. The video is about an hour long and Ken Schwaber gives a good and entertaining introduction to Scrum and talks about potential problems and how you can avoid them

15. Toyota – Toyota Production System:
    www.toyota.co.jp/en/vision/production_system/
    This Toyota website gives a good overview of the history of the Toyota Production System (TPS), and a very good introduction to the basic concepts of TPS. There is a small quiz you can use to test what you learned

16. Womack J, Jones D, Roos D (1990) The machine that changed the world: The story of Lean Production – Toyota's secret weapon in the global car wars that is now revolutionizing world industry. Rawson, New York

17. XProgramming:
    http://www.xprogramming.com/
    This is Ronald Jeffries' website about extreme programming with an introduction to extreme programming, a list of articles you can read online, and some links to software that supports software development using extreme programming

Chapter 4
Tooling

The previous chap. described agile software development as a set of principles and best practices that support a more flexible approach towards project management. The set of IT tools that are used needs to be able to deal with this kind of flexibility as well. Tools typically span software development, code version control system, test environment, project administration tools, and documentation tools. Which characteristics make these tools suitable for agile use?

4.1 Project Management Tools

In general, any good project management tool can be applied to any agile software development project. Key will be *how* the tool is applied.

Here is an example: the widely used software "Microsoft Project" allows you to efficiently create and manage a precise project plan, which assigns task to hundreds of developers. It can calculate milestones and durations based on sizing estimates and available team capacity. You can even optimize the plan by utilizing every single minute of each individual, while taking into account foreseeable circumstances such as vacation days. But such a plan brings us back to a key issue discussed earlier in this book: do not waste time with planning exercises on a too granular level of detail. The project status 6 months from now is not predictable!

Nevertheless, it can be absolutely useful to apply such a widely used tool that maps work to people and generates schedules and Gantt charts, as long as you use it wisely:

- Only make a plan for the short-term future (such as a single iteration).
- Limit the scope to reduce the planning complexity (e.g., each plan only focuses on a single team).
- Never ever blindly trust results generated by a tool.
- Revisit the baseline assumptions regularly (e.g., during a daily scrum meeting) and let the tool do the replanning.

T. Stober and U. Hansmann, *Agile Software Development*,
DOI 10.1007/978-3-540-70832-2_4, © Springer-Verlag Berlin Heidelberg 2010

In an agile world of self-organizing teams we are delegating a significant amount of responsibility to each team member. And we are emphasizing collaboration extensively: between peers, team members, stakeholders, as well as up and down the hierarchy. A local project plan on the hard disk of the project lead will not do it. Project management tools will be used concurrently by many users. They need to be used intuitively; they need to be distributed across teams, organizations, and systems. They need to be accessible to everyone to share remaining work to be done, as well as status and progress. Mixing the capabilities of project management and collaboration tools will create a new powerful tooling infrastructure for agile software development.

A review of existing project management tools shows that there a several categories:

- Standard tools, such as Microsoft Project, which are quite generic in use. They can be used for classical waterfall projects, but they can also cover some aspects of agile project management. Agile development and project management processes need to be implemented independently of such tools.
- Tools that are tailored to support a specific agile development process. Agilo for scrum is one example. The tool helps to manage the backlogs, scrum meetings, and also provides individual views for the various roles.
- Families of complementary tools such as Jazz support multiple aspects of the entire life-cycle of agile software development.

The following two sections describe two families of tools in more detail.

4.1.1 Microsoft Solutions Framework for Agile Development

Microsoft has extended its "Visual Studio Team System" to support concepts such as scrum, and to facilitate continuous integration of developed code.

The baseline of Microsoft's development methodology is a mapping of agile practices to corresponding functionality in Visual Studio. This mapping is referred to as Microsoft Solutions Framework for Agile Development (MSF for Agile). The methodology forms the foundation for the Microsoft Visual Studio Team System [5].

- *"Planning Game"* defines product and sprint backlogs using Excel and imports that data as work items into Visual Studio.
- *"Small Releases"* manages burndown charts. A scrum-like process can be defined as a "management process" in Visual Studio.
- *"Simple Design"* assists in the creation of technical design documentation leveraging "Application Designer" within Visual Studio. The designs and models are UML-based.
- *"Testing"*-tools within Visual Studio create test automation for unit tests and determine test coverage.

- *"Refactoring"* supports renaming existing code and transforming code based on parameters. Background is the Extreme Programming technique to continuously refine design and code.
- *"Collective Ownership"* comprises version control, work item breakdown, and tracking of progress. This functionality is covered by the "Team Explorer" tool.
- *"Continuous Integration"* is an automated build environment, which is integrated with the version control system. The build environment assists in integrating code changes quickly and executing unit tests on the code base continuously. Nice features such as automated notifications in case of build breaks are a tremendous help when dealing with self-organizing teams working in parallel.

The big advantage of the Microsoft tool family is that it is already a consistent part of the Visual Studio team system. Specific agile techniques such as scrum can be implemented leveraging the tool set, plus adding scrum-specific management techniques, like the Scrum meeting and the concepts of specific backlogs.

4.1.2   Jazz and Rational Team Concert (RTC)

Jazz is an open technology initiative aimed at improving the way people work together within a development project [1]. Jazz provides an extensible architecture and software platform to manage agile projects.

The IBM Rational organization, in collaboration with IBM Research, has leveraged the Jazz technology and goes beyond current agile project management solutions to include agile concepts such as team collaboration, transparent development, continuous integration, and test-driven development as first-class citizens. The resulting product is Rational Team Concert (RTC). It delivers an integrated project management environment that addresses the needs of various project participants and stakeholders.

Teams and project leads can manage the overall product backlog of a release, as well as specific work planned for an iteration (iteration backlog). Work items and user stories can be drafted and prioritized. Bugs and code changes can be triaged. Individual reports can be assembled to monitor development progress, e.g., based on burndown charts. A variety of dashboards can be composed from flexible building blocks to provide a quick, consolidated overview of the ongoing activities. Real-time reporting and alerts, in combination with a variety of metrics measuring velocity, test results, quality, and progress, make a project's status and risks visible. Educated managers will appreciate the value of these easily accessible dashboards, which are much easier to digest than multi-page Gantt charts.

Individual developers can pick up work items from the backlog, provide effort estimates, and report back on their progress and completion. By having the community members themselves enter the data such as work items and progress information, and by making the gathered information easily available to everyone

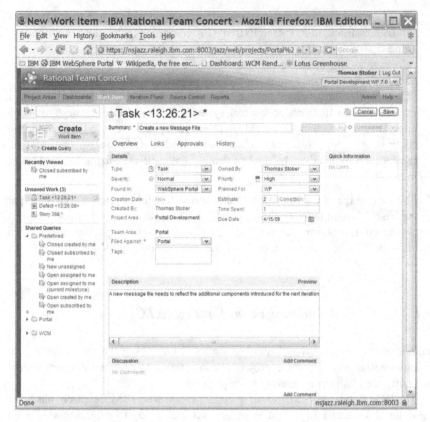

Fig. 4.1 Rational team concert provides a powerful web user interface to enter various kinds of project artifacts, such as tasks or use cases

in the project, Jazz is evangelizing a strong decentralization of planning and monitoring. This is clearly in contrast to having a centralized project manager who is the sole owner of the "plan."

Figs. 4.1 and 4.2 give an example, which illustrates the creation of tasks and the display of progress reports.

Rational Team Concert is designed as a platform where all project members can participate and to which all can contribute. As a result, the most innovative strength of the platform is the tight integration of collaborative features: Jazz technology provides team awareness by allowing team members to see who is present and what they are working on. Instant messaging enables collaboration in the context of the work currently being done. Definitions of teams and communities, discussion forums, as well as joint ownership of shared documents are just a few examples of team collaboration. This is nicely reflected by the fact that the generic term for any kind of document or artifact within the common repository of Rational Team

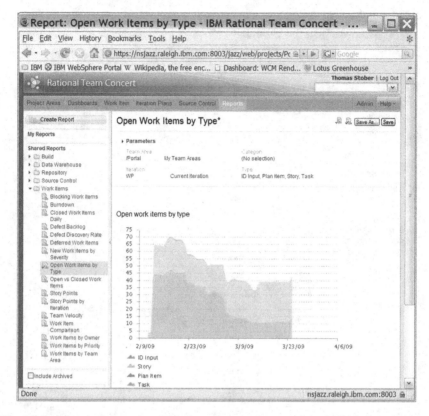

Fig. 4.2 Based on the information within the common project repository, individual status reports can be generated

Concert is referred to as a "team artifact." The goal is to leverage technology to create a cohesive team, even if team members are miles or even continents apart.

It is also notable that the common infrastructure of Rational Team Concert can be accessed either by an Eclipsed-based client application or by a browser-based web interface. Fig. 4.3 shows the user interface of the Eclipse application.

Rational Team Concert is also process-aware. The project environment can be customized to support different processes. There are several default templates for common processes such as Rational Unified Process (RUP), Eclipse development process or the lightweight scrum approach. These can be applied out of the box. In addition, virtually any project artifact, participant role and privilege, or process step within a workflow can be tailored individually to the project's philosophy and particular needs.

Rational Team Concert is not a single isolated project management product, but part of a rich portfolio of Rational tools. For instance, the Rational source control system ClearCase is integrated into the project management infrastructure and allows to couple the defined tasks directly with the code that implements them.

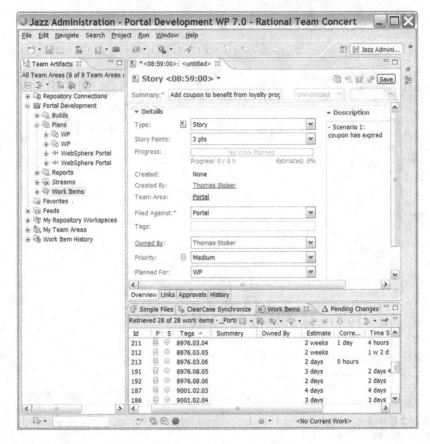

Fig. 4.3 In addition to the browser-based web interface, Rational team concert exposes its capabilities through an Eclipse-based client application

Build execution and regression testing can be triggered by Jazz as well. This provides a comprehensive tooling coverage from initial design and coding through testing and delivery.

4.2 Collaboration Tools

We have stressed the fact that any kind of agile software development will imply a substantial amount of communication and collaboration among peers, between teams, as well as up and down the hierarchy of an enterprise. This is especially true for large projects with teams spread across several locations.

Collaborative tools resulting from the ideas of Web 2.0 will be of invaluable help when facilitating interaction between individuals and teams. Web 2.0 comprises a

variety of technologies which especially make the web-based applications significantly more efficient to use and easier to deploy and manage. But most importantly,
Web 2.0 evangelizes a few fundamental nontechnical concepts:

- *Participation*: The community of users equally participates in contributing
  content to be shared and adding value for the greater good of the project.
  Information gathering and distribution is highly decentralized. There is no gate
  keeper for information repositories.
- *Reputation*: Despite the fact that team members contribute equally, there will
  always be users who are "more equal": They might hold distinct roles, such as a
  subject matter expert or a team leader. But individuals will also receive recognition via the visible contributions they add to the project.
- *Folksonomy*: The taxonomy and structure of the information repository is not set
  up in advance by an expert. It rather evolves by collaborative tagging, added by
  the consumers of the content. Artifacts and documents can be found based on
  evolved usage patterns.
- *Syndication*: Project-related information is not gathered in a single repository,
  but rather aggregated from various, decentralized, and independent sources.
  From these sources, a consolidated view is produced by an overarching portal,
  which provides the illusion of a nicely integrated system to the user.

These concepts are extremely helpful when applied to agile development projects and
the tooling infrastructure used. Collaborative project management platforms such as
Jazz and Rational Team Concert clearly promote and implement these concepts.

But even without such sophisticated, next generation project management tools,
collaboration can be substantially improved using simple means: generic Web 2.0
applications, such as blogs, wikis, instant messaging, web conferences, or community platforms can provide considerable value for jointly planned iterations, evolving of designs or elaboration of end-user documentation.

In this context, especially an ordinary wiki turns out to be an extremely simple,
but very powerful tool that allows to easily involve a distributed community of
internal or even external experts.

Nice examples of comprehensive collaboration and communication products are
Lotus Connections and Lotus Quickr, which combine Web 2.0 capabilities with a
consistent user experience. A sample of Lotus Connections is shown in the following Fig. 4.4.

4.3 Development Infrastructure and Environment

The choice of the ideal development environment will not make a significant
difference. It is more along the lines of personal belief. Some will love a simple
editor like SlickEdit. Others will prefer an integrated development environment
such as Eclipse or Rational Application Developer.

Fig. 4.4 Lotus connections exposes a dashboard for collaboration, activities, and communities

Whatever the choice might be: not the development environment itself, but the underlying infrastructure is the key to efficiency and success.

4.3.1 Source Control and Version Management

First of all, each developer must be able to easily access the most recent code base. This makes a flexible and powerful source control system like ClearCase very important. Source control systems manage versions and changes of source files as well as specific development features and detected bugs. Several development and project management tools, like Rational Team Concert, can directly and seamlessly integrate source control and version system.

4.3.1.1 Build Infrastructure

In agile projects developers will continuously integrate their code changes into a common code base. The production build will constantly pick up the changes and produce updated versions of the deliverable over and over again. A fast build turn-around is an important step to enable developers to implement changes or fix bugs and allow them to do a full verification of their code change immediately. This makes the build tooling and infrastructure another essential piece of the development environment. Regardless of whether the build tools of choice are based on Ant, Maven, Rational BuildForge, or MSBuild, they need to generate binaries, documentation, product installer, and distribution media, and all with a single command.

Fast build machines and optimization of the build setup and tools are absolutely worth the investment. Ideally, a complete product build should be done in less than 15 min to provide quick feedback to the programmers. Unfortunately, this will not always be achievable in larger projects.

The second-best approach is to enable developers to run a copy of the product build in their own development environment and to enable them to just compile their single change locally. Afterwards, they can run their part of the test suite to ensure that their code change works at least in their local environment based on the last published version of the production build.

But this approach adds the risk of version conflicts when developing concurrently. There can also be inconsistencies between local builds run in parallel by different developers.

If a longer build duration or less frequent build execution is unavoidable, a workaround can be to introduce some code verification prior to the production build. For example, you can implement several shorter builds that each cover a part of the daily product build and are automatically executed sometime during the day to minimize the risk of a problem during the product build Fig. 4.5.

The first step that a code change must pass could be a checkpoint build. The checkpoint builds are executed frequently for each individual component. They can run in parallel and have the purpose to ensure that there are no basic compile problems with changed code. Only code changes which successfully passed a checkpoint build are promoted to the next level: the prebuild.

The subsequent prebuild is a copy of the full production build to identify any build breaks early. As before, only code that passes the prebuild successfully is promoted to the final level, the production build. The build system sends automatic

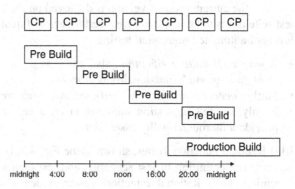

Fig. 4.5 Daily build schedule with checkpoint builds, prebuilds, and production build

emails to the involved developers if any code problems are experienced during checkpoint build or prebuild.

Production builds are executed once a day and produce customer-ready installation CDs as well as images, which can be directly copied onto test machines.

The goal of the whole infrastructure is to make sure that problems are identified long before the production build, ideally in the developer's own test environment, even before they put the code into the version control system.

A flexible and scalable software architecture that has clearly identified components with agreed-on services and interfaces will also help to minimize the implicit dependencies between the different parts of the product and reduces integration problems.

4.3.2 Automated Test Environment

Last but not least: Developers need to be able to get real-time feedback for each single code change immediately. There will be continuous regression testing of the build deliverables. This implies that the development infrastructure includes tools to quickly set up test machines and to trigger a suite of regression tests automatically. A fast turnaround, when running the regression testing suite is crucial.

Installable images for each target platform can be produced directly by the production build process. In the simplest case, each developer can easily copy installed product images directly to his development machine. RSync [3] is an open-source tool to synchronize two file systems and turned out to be useful in this context.

Setting up test machines will become a tedious endeavor as soon as the deliverables imply a complex system environment, such as server clusters or several different software platforms. If the complexity grows, additional tooling becomes crucial in order to limit the costs of test execution. Automated deployment tools can be applied to set up, monitor, and manage all pieces of a comprehensive test environment. Rational Build Forge and Tivoli System Automation (shown in the Fig. 4.6) are two examples for this category of tools.

Once the current product version is deployed on the test systems, the regression test suite needs to be run to ensure a certain level of quality. There will be multiple levels of automated regression testing:

- A very brief *build verification*, also know as the smoke test, will check if the most basic product functionality is working.
- Further *acceptance tests* will verify the major features of the product.
- Finally, the *full regression* suite will cover a very broad set of test cases and provide a thorough quality assessment.

IBM Rational Function Tester, shown in the Fig. 4.7, is one example of a tool that supports the definition of test scenarios and test data and drives their automated execution. With Rational Function Tester, a developer can either record and

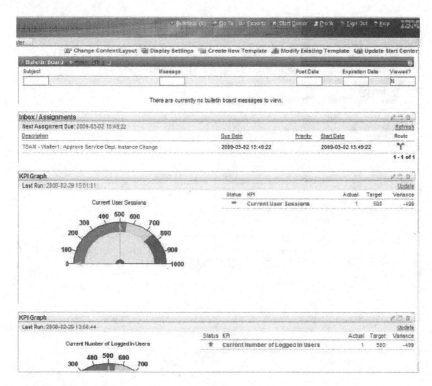

Fig. 4.6 Tivoli Systems Automation for Multiplatform (TSAM) is one example of a tool to set up, manage, and monitor a complex testing environment

generate functional testing scenarios, or write testing scripts. During execution the tool will compare actual with expected results and produce a report of the quality status for each tested build.

4.3.3 "Code-Build-Test"

Code-Build-Test is the basic cycle, which makes up the pace of the project. Efficient and seamless integration of the tools involved in this loop has a more relevant impact on the overall turnaround speed, than having more or less features within a particular development tool.

Three aspects are extremely important in this context:

- *Local Testing*: Changes must be validated locally before they are checked into the source control system and become part of the code base. This must be an easy-to-do task trigged by a single command and must be a self-evident thing to do for everyone involved in the testing.

Fig. 4.7 Rational Function Tester is a tool to create and trigger automated regression tests

- *Duration*: Adding and integrating the pretested changes to the code base implies a recompilation within a production build and some level of regression testing. The duration of this process will be crucial. Optimizing this infrastructure is vital, as it is extremely painful and inefficient to wait for results of a code change. In large projects an instant turnaround might be difficult to achieve. Smart componentization can lead to some parallelization. Running the production build on a high-end machine can accelerate this process to some extent as well.
- *Ease-of-Use*: If the process of coding, building, and testing is too cumbersome, developers may be tempted to take shortcuts and take the risk of adding untested code to the project. A nicely integrated infrastructure of development environment, source control, local and production build process, as well as test tooling can help minimize this risk and will simplify life tremendously.

Further Readings

1. Jazz Community Site:
   https://jazz.net/
2. Barnett, Liz: The IBM Rational Jazz Strategy for Collaborative Application Lifecycle Management
   http://www-01.ibm.com/software/rational/jazz/analystWP/
3. Rsync:
   http://rsync.samba.org/
4. IBM Rational Tools:
   http://www-01.ibm.com/software/rational/
5. Microsoft Solutions Framework for Agile Development:
   www.microsoft.com

Chapter 5
Considerations on Teaming and Leadership

So far we have discussed a vast range of best practices, techniques, guidelines, and tools that are typically associated with agile thinking. By now, one important question is likely to come up: How can we apply these to a real-life project? What are the considerations that need to be taken into account? In the next three chap. we will look at specific aspects that need to be considered in order to make your agile software development project successful.

We have structured them into three categories:

- Teaming and leadership (Chap. 5)
- Architecture, design, and planning (Chap. 6)
- Execution, implementation, and test (Chap. 7)

5.1   A "Lean" Hierarchy

So far we have learned that speed is one of the prerequisites for success. We want a quick turnaround of product cycles, which reflects the current needs of the market. We want our development to adapt to change and continuously optimize its processes as well as the products. We want motivated and innovative teams, with the spirit to perform and deliver customer value without being hindered by confinements of regulations. We want to foster a climate of smart collaboration and intensive communication to bring the best skills together and to come up with interdisciplinary creativity.

In the IT industry, a competitive advantage can be measured in weeks, not in years. Change is abundant. The ability to use change as a driving force to gain competitive advantage is key to economic success. Organizations that are hierarchically organized and have been stable for many years will most likely be unable to cope with change.

T. Stober and U. Hansmann, *Agile Software Development*,
DOI 10.1007/978-3-540-70832-2_5, © Springer-Verlag Berlin Heidelberg 2010

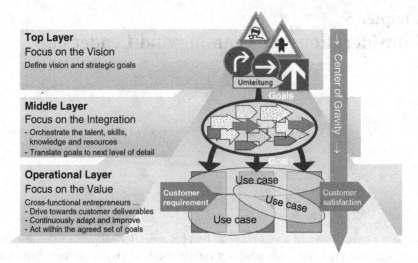

Fig. 5.1 Layers of an organization

What can the hierarchy of a "lean" and adaptable organization look like? Strongly simplified, such a hierarchy within a company can be split into three generic layers that differ in their responsibilities and purposes (see Fig. 5.1):

- *Focus on the Vision:* The top layer defines the vision and strategic goals. Crucial areas of investments and directions to be pursued are decided here and are made transparent throughout the company. This level combines representatives of executive management, senior technical leaders, as well as customer relationship representatives such as sales and marketing. They own the product and sponsor the project.

- *Focus on the Integration:* The second layer comprises middle management and project managers as well as software architects. Their responsibility is to orchestrate the talent, skills, knowledge, and resources of the company to work towards the high-level goals. For this purpose, the overall goals need to be translated into more precise goals of individual teams and projects. The challenge is to ensure that the full set of goals for all teams are consistent and without contradiction. The deliverables of each team need to contribute to these goals. It is obvious that these sets of goals will be dynamic and will adjust whenever constraints and requirements change.

- *Focus on the Value:* The operational layer is responsible for the creation of value. This layer is structured in multiple cross-functional, interdisciplinary teams. Each team is accountable for a set of use cases, which are implemented end-to-end and deliver direct customer value. Each team owns the responsibility and authority to pursue its goals within the scope agreed with management. Most importantly, each team will continuously adapt its goals, use cases, and deliverables to reflect any changing constraints.

It would be ideal if there were no dependencies between different teams in one project, but this is usually not the case. Different teams are interconnected among each other through customer-supplier relationships as well as to the real customer. These relationships help to ensure a direct link between individual team members and their customers. They will also provide a clear understanding of customer needs. An important generalization is that customers can be external as well as internal ones. Being accountable for the delivery of customer value makes each team member an entrepreneur. This lowers the center of gravity within an organization.

It is notable that the vertical integration of the company is achieved through a system of goals which is detailed from the top down to the operational layer and ensures that all teams are aligned and work towards a common vision. The established customer-supplier relationships between the individual teams create a horizontal integration and ensure the flow of communication between the development activities that are pursued in parallel.

By combining the goal-driven and the customer driven-view, and by emphasizing teams that take over responsibility towards goals and customers, a powerful dynamic is triggered within an organization. Fig. 5.2 shows these different kinds of integration.

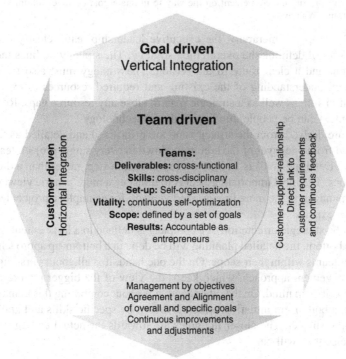

Fig. 5.2 Goal-driven, customer-driven, and team-driven view

Rapidly building quality software relies on teaming. Successful teaming relies on a mindset in which communication and relationships are more relevant than documentation and formal processes.

5.2   Setting the Directions: Management by Objectives

We keep emphasizing goals as the integrating link between different layers in a hierarchy. Establishing well-defined and consistent goals becomes significantly more important as soon as individual teams begin to develop more autonomy. A decentralized organization will only be able to make progress toward a common greater good, if the scope of each team is clearly defined and in line with a corporate vision and strategy. Agile software development makes the role of leadership much more demanding, despite the fact that we are suggesting the delegation of a significant amount of responsibility to the teams and individuals. Leadership by defining strategic goals with an appropriate level of abstraction, rather then following the temptations of micromanagement, requires a distinct visionary talent. Leading and coaching instead of directing means to let the people closest to the object of a decision make that decision.

> Don't be a time manager, be a priority manager. Cut your major goals into bite-sized pieces. Each small priority or requirement on the way to ultimate goal becomes a mini goal in itself. (Denis Waitley)

The top layer of the hierarchy, the executive leadership team, clearly owns the responsibility of defining the overarching strategy. The strategy outlines the long-term vision, and a clear route to that vision. The strategy must also include an agreed-upon understanding of the existing and required resources to execute towards that vision, as well as a strategic plan to close any resource gaps. Resources in this context can be people, funding, skills, or technology.

Goals are derived from the strategy and are prioritized and detailed as they are propagated from one layer of the hierarchy down to every single project, team, and individual. Detailing goals is an interactive translation step, rather than a dictate of expectations. The most important aspect is to translate a more generic view, such as that of a manager, to the more fine-grained view of, for example, a project leader or a technical experts.

While the strategic orientation of a company is defined to a large extent from the top to the bottom, the detailed planning will be done in a bottom-up approach by the individual teams within their scope. On the one hand, it is all about pursuing a top-down management approach, which keeps the view of the bigger picture and the overall priorities in mind. On the other hand, it is about combining this management view with a bottom-up initiative, which is driven by specific skills and knowledge of a specific subject. The closer the goals are towards the actual coding work the more precise they will be.

> First you write down your goal; your second job is to break down your goal into a series of steps, beginning with steps which are absurdly easy. (Fitzhugh Dodson)

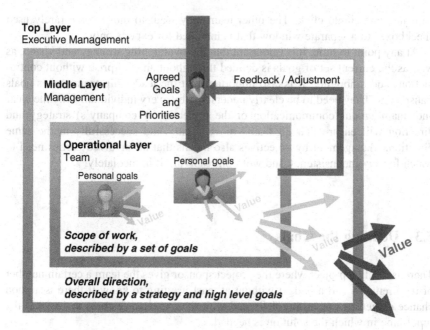

Fig. 5.3 Prioritized goals define the scope in which a team can act

Agreed goals will spell out *what* needs to be done, but they will not detail *how* things will be implemented. Instead, they will outline the priorities and the scope within which a team is allowed to execute. The goals will make clear in which way a particular team is expected to contribute to the greater good. Along with a set of goals, a team will inherit the necessary responsibility and authority to pursue suitable tasks and work items on its own. Empower the teams! Let exactly those people drive fine-grained decisions who know the details first-hand and will have to implement what they have decided on. Managers can not expect teams to perform at a high level without allowing them to be the authors of the solution.

Obviously, the better the personal goals of an individual are aligned with the corporate or project-specific goals, the more powerful will be the performance of the entire team (Fig. 5.3).

The following example shows how goals could be propagated within a company: Assume that part of a corporate strategy is "Helping clients to be more successful by making products easier to consume." The derived company's high-level goals might be to enhance all its software products to better address the needs of unskilled end-users. Based on these directions, a specific project may decide to achieve these goals by simplifying the install utility and by also revamping the user interface of the software. The project leader will appoint two teams: one team will focus on the install utility, while the other team will work on redesigning the user interface. Each team will bring the goals to the next level of detail, e.g. the install team might come to the conclusion that it needs to enhance the utility with an option to install with just one single click. The other team may want to move some rarely used checkboxes to a separate window that is intended for experts only.

At any point in time, it is important that the overarching strategy and vision, as well as the current set of goals is defined throughout an enterprise without contradictions and ambiguity. It is also important that the leadership team makes goals transparent. They need to be clearly understood by every individual. Only the clear and unambiguous communication of the project's (or company's) strategy and direction will ensure that all teams move jointly and successfully in the same direction. Management by objectives also means that all involved parties need to watch for any inconsistency and work on resolving it immediately.

5.3 Defining the Goals

There is rarely a project where the project sponsor gives the team a certain number of requirements and a budget without at least wanting to know if there is a good chance of getting something delivered with the expected features and within the timeframe in which the solution is needed.

This is even more important in service projects, especially external ones, as here the customer usually asks several companies to make offers and then chooses the one that delivers what he is asking for in the required time, quality, and for the lowest cost. Unlike the buying of nails or screws, the commission of a software projects requires the customer to have a certain level of confidence that the bidder chosen for the contract is capable of doing the job and will also be around later to service the result of the work. In addition, the project sponsor is faced with the challenge of assessing the quality of a future solution which the bidder will be not be delivering until a few months later.

The effort estimates before the start of a product development can only be very rough. They are more precise if the team has already done similar projects in the same skill domain and was using the same technology. The risk is higher if the team has or wants to use new technology (such as AJAX today) that is still in the early days of adoption.

If a binding offer needs to be made based on rough estimates that the team has put together, the need arises to assess which risk one wants to take. One can play safe and make a high-priced offer which contains a comfortable amount of buffer for unforeseen surprises. Alternatively, one can be aggressive, assume that the estimates are exact and make an offer without calculating additional buffers. A third option would be to choose a compromise between the safe and the aggressive approach by adding a bit of a buffer. With the safe option, you will be able to make some money on the project, but the chances that you will even get it are low. With the aggressive option, chances are high that you will get the project, but chances are also high that you will lose money on it. A tough decision – but it is unlikely that you will minimize the risk by trying to make a very precise estimate and spending

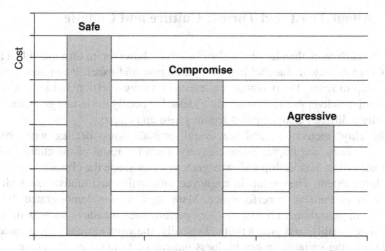

Fig. 5.4 Different more or less aggressive offers one can make while bidding for a project

a lot of time on the estimating and project planning (and possibly still not get the project in the end, in which case you will definitely have lost money on it) (Fig. 5.4).

The situation is often easier for an internal project. The project sponsor will still want to know what he will approximately get and when. But in this case you may be able to convince him to agree on high-level goals and let the team work out the details over time, with regular checkpoints once a week or at the end of each iteration.

In this case some rough estimates at the beginning of the project might be good enough to get the project going. It is important that potential dependencies between different teams are identified up front, but spending too much effort on detailed estimates can quickly become a waste of time and effort. In the end, reality will always differ from what you planned. One of the key ideas of agile thinking is to accept this as a constraint for software development projects and use efficient ways to deal with this risk. It is important that the complete team is involved in setting the goals for the project and that there is buy-in and support from them that these goals are achievable. These high-level goals should give the team room and flexibility when working with the project sponsor on the lower-level user stories. It all depends on the level of trust between the project sponsor and the team. Of course there needs to be a level of project management that makes the project predictable and gives the project sponsor an early heads-up in case something goes wrong. The worst thing that could happen is that the project sponsor is unpleasantly surprised at the end of the project. Which again stresses the fact that customer and project sponsor involvement is key to a successfully project. Depending on your project, the project sponsor may not be the customer using your product, as in the case of commercial software product development. In these cases, it will be your product management, sales team, and the executive team which acts as the customer.

5.4   About Trust and Thrust: Culture and Climate

It takes much more than leadership by objectives, however, to turn individual team members into actual shareholders who participate and excel. It requires a strong leadership to reveal the potential that exists in an organization and to turn it into competitive deliverables and innovative ideas. Especially the letting go of control is typically a difficult challenge that requires trust and courage.

Mentality, motivation, and the overall attitude when dealing with change, success, failure, and improvements, is very much the result of the climate within an organization. Leadership and management styles shape the climate at work to a very large extent. They result in employee satisfaction and motivation, which in return drives the team's performance. Many studies have demonstrated the link between organizational climate and key performance measures such as development time, quality, and productivity. Typically, the work climate is accounted for 10–25% of the variance in key business indicators. In many cases, a company or project lacks the awareness for the change needed in its prevalent mentality.

A 2004 survey by TNS showed that within the United States about 40% of workers feel disconnected from their employers and two out of every three workers do not identify with their employer's business goals and objectives. In 2008, the market researcher Gallup estimated the damage to Germany's economy through those employees with only little or no emotional relationship to their employer at approximately 250 Billion Euro each year. This is roughly the budget of Germany's entire government. A team has missed to tap a significant resource if an employee excels at home by coordinating the fund-raising event of the local soccer club, dedicating his heart and soul to the task, but will not dare take over responsibilities in his day job. A workgroup cannot afford to loose talent – talent is the ultimate source of value!

A corporate culture can give some guidance on the type of mentality that is desired. A corporate culture is defined by the executive leadership team and can spell out specific values that support the establishment of a positive work climate. A culture can be a framework that imposes standards of behavior that specifically reflect the objectives of the organization.

As an example, IBM's culture spells out three distinct values:

- Dedication to every client's success.
- Innovation that matters – for our company and for the world.
- Trust and personal responsibility in all relationships.

Sam Palmisano, IBM Chairman, President and Chief Executive Office, describes the purpose of IBM's culture like this:

> If there's no way to optimize IBM through organizational structure or by management dictate, you have to empower people while ensuring that they're making the right calls the right way. And by 'right,' I'm not talking about ethics and legal compliance alone; those are table stakes. I'm talking about decisions that support and give life to IBM's strategy and brand, decisions that shape a culture. That's why values, for us, aren't soft. They're the basis of what we do, our mission as a company. You've got to create a management system

that empowers people and provides a basis for decision making that is consistent with who we are at IBM. (Sam Palmisano)

In addition to the overall corporate culture there will always be specific standards of behavior within individual teams. Each workgroup will cultivate its own behavioral peculiarities and their style of interaction. The climate within these microcosms might deviate from the corporate one for good or bad. It will be strongly influenced by the team's leader and it will affect the whole system!

Trustfulness is one key precondition for a positive climate within a team, as trust is the most important source of thrust. This includes an open and honest communication. It is no coincidence that IBM explicitly lists "earning trust" as a leadership competency. We are not suggesting that leaders ought to hug, award, praise, or pamper their teams and generously distribute T-shirts showing their project logo. We are simply emphasizing honest involvement and respect!

Inevitably things will go wrong occasionally, and a team will need to adapt yet again. While a workgroup is accountable for its results as a whole, failure is not the trigger for blame and punishment, but rather a source for initiating improvements. It can never be the other one's fault, because everyone is in it together. Failure or bad results are not bad, but a trigger for discussion. Agility is expediting improvement, not persisting in one's judgments. Björn Bloching, Partner of Roland Berger Consulting, underlines this with this statement:

An organization, which does not tolerate mistakes, cannot innovate, since everyone is only focusing on securing himself

The organization will trust the team to recover from the inevitable mistakes happening along the way. Trust means people can admit to their mistakes without fearing retribution, poor performance reviews, or even loss of their status. Nurturing such trust requires time and effort. Team members will learn to accept feedback as input from peers, rather than as criticism of rivals. It is important to implement a culture that is able to deal with success as well as failure, e.g. by having lessons-learned sessions or doing reflections on our work. Both the development process and the developed products need to adapt quickly, based on experiences made.

A corporate culture and a team climate with a positive attitude towards failure will lay the cornerstone for high performance within a workgroup. Thomas John Watson (1874–1956), chairman of IBM, solaces:

"You can be discouraged by failure, or you can learn from it. So go ahead and make mistakes, make all you can. Because, remember that's where you'll find success – on the far side of failure. If you want to increase your success rate, double your failure rate. (Thomas John Watson, Sr.)

5.5 Cross-functional Teams

We have learned that a lean management will surrender a significant amount of responsibility and authority to the individual teams. With a lower center of gravity, an organization can act and react quicker and be more focused on the customer needs, as the teams are directly in touch with their customers. But will autonomous teams, which adapt and organize their microcosm on their own account, lead to anarchy? We have seen that management by objectives is one instrument to give a set of independent, bottom-up initiated efforts a desired common direction.

The way teams are set up within an organization is also of relevant in this context.

In an earlier chap., we mentioned the work of Frederic Winslow Taylor. When introducing the division of labor, he suggested splitting the planning aspect of work from its execution.

> The task is always so regulated that the man who is well suited to his job will thrive while working at this rate during a long term of years and grow happier and more prosperous, instead of being overworked. (Frederic Winslow Taylor [6])

While this helped to improve the profitability of large-scale mass production, it had several significant downsides: rules set by the planners diminished the vitality and ability of the workers to create innovation. Furthermore, this approach implied a strongly functionally oriented split of work. Employees were fulfilling their duty without understanding the context of their work. Most of all, employees were shielded from the customers and their particular needs. Instead of providing direct exchange between supplier and customer, communication was anonymous and distorted by passing through an organization's hierarchy.

While an isolated function can be optimized to its perfection, is it difficult to get an overarching assessment of the overall workflow and the produced value in its entirety. A function-oriented split of work tends to increase the necessary investment for communication and logistics, due to many interfaces between the individual functional units. Coordinating and controlling functional units and maintaining a reliable flow of information through many interfaces will quickly translate into significant costs. A reliable communication across the units will be hard to establish and maintain. Integrating deliverables from multiple units to the greater good is often a tedious endeavor with a significant risk of failure.

This approach does not scale when the comprehensiveness and complexity of the new, unique project and its deliverables skyrocket.

To maintain a functional focus, an alternative is to entrust individual teams with entire end-to-end tasks. Breach the functional silos! This implies moving towards a horizontal view of produced value instead of cutting a project into a functionally oriented structure. An interdisciplinary team assumes the ownership and leadership for a particular part of a customer deliverable. Within a software development project, a specific team could be asked to implement a certain use case, starting from its initial design up to the final regression tests. On this journey, the team will touch multiple functional areas and add code to different components. A team will create user-valued capabilities rather than focusing on an individual layer of a component of the product's software stack. The idea is not to have separate teams develop the database layer, the user interface layer, and so on.

One consequence is a change in the code ownership: Teams or individuals will no longer own particular code or components. Instead, the ownership is shared and teams will adapt various parts of the code base in order to fulfill their mission.

The team will have a holistic view of the scope of work and will cover different disciplines such as programming, testing, performance measurement, or documentation. They will work in parallel with other teams that will be responsible for other use cases. The team will also involve customers directly and balance customer needs with possible solutions.

It is exactly this kind of direct interaction that will establish sustainable supplier-customer relationships between teams. It will weave strong connecting links into an organization and will establish a consistent horizontal integration throughout the company, as we've advocated earlier. When pursuing end-to-end use cases without many hand-over or middlemen, significantly less logistics and coordination will be necessary. In the final analysis, this will prove to be more efficient and less error-prone than putting together a deliverable from separate functional blocks developed by specialists, and need to be integrated afterwards.

Even virtual teams can benefit from this approach: It almost goes without saying, that sitting next to each other is definitely the most effective way to collaborate and should always be preferred, when setting up teams. But it is often unavoidable to recruit project members from multiple locations. In this case, cross-functional, virtual teaming can help building powerful relationships by focusing on a common agenda rather than just handing over components between individual locations while at the same time driving ones own local interests.

In our earlier example, the team that modifies the user interface will add a new component managing the checkboxes for the expert mode. But they will also touch the code that was written for all other windows in order to simplify them, regardless of the original code ownership. The team will consist of a few developers, testers, a software architect, a user interface designer, and a technical writer.

Some tasks might not be fulfilled in the most efficient and most perfect way, as the team might not have the ultimate expertise for all concerns. In fact, the presence of the perfect set of skills within a team will be rather unlikely. But there will be less hand-over between different teams, less interfaces to coordinate, and less puzzle pieces to integrate afterwards. Within a team, everyone needs to know and understand what the other ones do. Each individual can make a difference, and everyone is in it together.

Such a view can be thrilling and will boost effectiveness, with teams adopting a holistic end-to-end view of their mission.

5.6 The Wisdom of Crowds

When Frederic Winslow Taylor propagated the division of labor in order to achieve higher productivity, he did not leave the individual worker significant scope to excel or think. According to Taylor, each individual will optimize their set of skills on their particular task. Their destiny will be to continue this local role without ever looking beyond the boundaries of their cube or even challenging their surroundings. With an increasing amount of communication, logistics, and continues change of projects and products, this isolated view of a job description cannot be sustained.

It is amazing what can happen when team members start to interact and organize their local microcosm as a work group.

There are many impressive examples where innovation is triggered by individuals or groups even without official assignment of the responsible management. This kind of bottom-up, nonprogrammed activity for the benefit of the company is referred to as bootlegging. Bootlegging activities are not included in any project activity, nor are any formal resources allocated towards it.

Nevertheless, some companies explicitly support bootlegging. For instance, Google grants 20% of an employee's work time to personal projects related to the company's business.

A famous example of a bootlegging activity is the yellow sticky Post-it note: The inventor of an odd sticky adhesive substance, a 3M employee and chemist named Spencer Silver, couldn't think of anything useful to do with it. On the contrary, the glue was considered a failure because it didn't stick very well. It took another colleague, Art Fry, to come up with a practical application: Tired of searching for the songs in his hymnbook at church, he applied the adhesive to anchor shreds of paper as bookmarks without permanently damaging the pages. Combining their unique skills, the inventor and the salesman morphed this idea into adhesive scratch notes. Despite the fact that 3M permits employees to spend 15% of their time on their own undertaking, it proved a strenuous venture to work their way up the management chain to promote their discovery. Eventually, after a long struggle for internal support and funding, the idea was productized as an easy-to-use communication tool. Post-it notes turned out to be a tremendous commercial success and 3M earned a fortune with this product.

However, this example also tells another truth:

The example does not only illustrate the fact that innovation in an organization is driven from the bottom to the top. It also becomes evident that combining two different skills can achieve what the same skills would not have done while working in isolation.

The ability to collaborate is what makes a team successful: Knowledge and skills are no longer just a competitive advantage of individual employees when they pursue their own career. Instead, knowledge and skills are a puzzle piece which can, when added to other pieces, shape something entirely new. Innovation soars when team members exchange ideas and benefit from mutual inspiration. Teams that master interdisciplinary collaboration will outperform isolated, technical expertise.

Even complex systems can arise from a multiplicity of relatively simple interactions or by rearrangement of already existing entities. In philosophy, systems theory, and science this phenomenon is referred to as emergence.

Emergence can be defined the arising of novel and coherent structures, patterns and properties during the process of self-organization in complex systems (Jeffrey Goldstein, 2002)

> Every resultant is either a sum or a difference of the co-operant forces. Further, every resultant is clearly traceable in its components, because these are homogeneous and commensurable. It is otherwise with emergents, when there is a co-operation of things of unlike kinds. The emergent is unlike its components insofar as these are incommensurable, and it cannot be reduced to their sum or their difference. (George Henry Lewes [3])

Collective intelligence is a shared or group intelligence that emerges from the collaboration and competition of many individuals. Collective intelligence is not limited to workgroups within a single organization. It is explicitly targeted at large virtual teams, such as open source projects.

> "Collective intelligence is the capacity of human communities to evolve towards higher order complexity and harmony, through such innovation mechanisms as differentiation and integration, competition and collaboration." (George Pór [5])

Characteristics of collective intelligence are:

- *Sharing*: Ideas are shared at early stages among team members as well as externally. Feedback as well as suggestions for improvements is solicited in order to accelerate product development.
- *Peering:* Knowledge exchange and development activities are self-organized among peers, in order to gain access to new ideas and technology.
- *Openess:* Intellectual property is exchanged liberally, even across the boundaries of a company, in order to seek new opportunities and reach out to new technology.
- *Acting Globally:* Virtual teams within a globally integrated company reach out beyond boundaries of organizational units in order to combine the best possible set of required skills.

Collective intelligence as well as any kind of teaming can excel when new forms of networking enabled communications technology are leveraged, as demonstrated by the emerging second generation Internet concepts of Web 2.0. In the world of Web 2.0, team members can interactively generate content, share information, and extend the pool of existing knowledge.

5.7   It Isn't that Easy

To be honest, the idea of teams taking a holistic end-to-end view of their mission sounds great, but it isn't that easy to implement. Acting as leaders is fairly demanding in an agile environment. Working within an agile development team means to face tough challenges as well.

First of all, the lack of guidance in many details of the daily work will be difficult, especially for junior developers without much experience in development projects. Especially in large projects, there is a risk that individuals can simply get "lost" in the crowd. Agile software development demands proactive initiative, self-motivation, and the ability to work independently.

This may in some cases be too much to ask of inexperienced teams. A possible mitigation can be to get started with a strong, supportive leadership and begin to hand over responsibility incrementally as those new to the game gain experience.

Often a minority of team members will generate the majority of contributions to the team and will pick up a kind of leadership role. But highly collaborative teams cannot afford a majority of people sitting back and waiting for others to take initiative. We are in it together! The active team members need to encourage the silent majority to abandon their restraints and become involved. The more eyes, brains, skills, and knowledge are applied to a tough problem, the more likely the team will succeed! Active mentoring is of great help in such a situation.

Despite the fact that joint collaboration produces the best results, "leadership" will move informally between team members, as appropriate. Different individuals will assume leadership for specific topics. In addition, everyone participates and picks up responsibility for the entire group (rather than limiting their view to the boundaries of their office cubical). Within this mindset, fewer things will fall through the cracks. Errors and oversights are less likely to occur.

Second, the increased need for collaborating puts a much stronger focus on soft skills, social intelligence, and communication abilities than on traditional technical expertise on specific subjects. While there will always be a need for a technical guru who has the ultimate wisdom on a certain piece of technology, the top performer of a company will be an all-rounder and good team player. This person will have a personality that brings together different interests, and understands how to assemble a solution out of various fragments and thoughts. He or she will think and act as an entrepreneur. Aligning goals, initiating work, resolving conflicts, and involving external customers directly will add to the qualification needed for the daily work. For instance, it is quite easy to identify clashing goals. It might also work rather quickly to get these resolved by making your manager decide one way or the other. But it will definitely require a lot experience in human interaction to work out a really good compromise that combines the best of each option, gets the buy-in of all involved parties, and results in a better solution.

This will be even more difficult in geographically distributed teams, where you need to establish a common team spirit across different locations.

To a large extent collaboration and teaming is still neglected in most education facilities, which still aim at transforming students into masters of technology and preparing them for a life as solitary gurus.

Sharing, peering, and open collaboration is often in conflict with the still predominant attitude that considers knowledge and skills as a competitive advantage of employees pursuing their own careers.

Focusing on the business problem rather than on personal career goals is often difficult. Especially the highest performers tend to force their views at all cost in support of their own reputation and personal recognition, or they just run off to solve a problem on their own. Collaborative team members, who are assessed primarily on the results the team produces as a whole, will attempt to leverage the skills of the entire group rather than pursuing their own personal agenda. As a matter of fact, teams become more efficient and successful once they achieve a higher level of collaboration to accomplish the common goals, instead of counting on individual heroes saving the day.

Measuring how much value an individual provides to the team is difficult. The most valuable team players often have lower individual productivity, since they invest a significant portion of their time supporting and mentoring other team members.

Encouraging collaboration and teaming can involve many different measures:

- A working environment for open communication flow is one aspect. Often labs and offices are designed machine-friendly instead of human-friendly. A collaborative working environment is built around open spaces, not around cubicles.
- If you are colocated, talk directly and face-to-face. Some issues can be solved in the most creative way while holding a cup of coffee. Invest time in bringing the team together and celebrate events, like exiting an iteration. Build relationships!
- Distributed teams need to figure out a way to work around geographical distances. Virtual workspaces, Wikis, instant messaging, and web conferences work very well and scale. But the direct and frequent contacts with your colleagues are even more important if individuals are not sitting next to each other.
- Peer-to-peer ad-hoc communication will clearly show better results than waiting for the next scheduled meeting or sending emails to large distribution lists and waiting for replies. Get involved! People need to interact, negotiate, talk, phone, chat, discuss, rather than just handing over work items to someone else.
- Let teams focus on their work! Limit distraction by side jobs and avoid team members being exchanged too often between different projects.

These are just a few measures that can help let a common team spirit grow. A strong feeling of "we" and a pride of ownership in what the team delivers is a powerful asset within a project!

> People fail to get along because they fear each other; they fear each other because they don't know each other; they don't know each other because they have not communicated with each other. (Martin Luther King Jr)

When moving an organization towards an agile and lean management approach, there will be a lot of radical and far-reaching changes that will unsettle people. Indeed, there will likely be a lot of opposition towards introduction of agile methodology. Keep in mind that agile development is more than just a set of programming tools or best practices: agility reaches out to your entire organizational setup, it will change the work climate, and it challenges roles and responsibilities. Familiar structures will go away and transitioning will be disruptive. Teams will be inaugurated and disbanded as needed. Job responsibilities will change much more frequently. There will be fear that the current familiar work assignments will go away. And the ambiguity about future work assignments beyond the horizon of a few weeks will make people somewhat nervous.

Human life is just an instance, its existence is subject to continuous change. (Marc Aurel)

To better accept these concerns, it is helpful to understand two fundamental behavioral patterns as outlined by Edward T Hall [1]:

• *Monochronism* demands a planned, deliberate control over time. This is a task-oriented way of living. Individuals like to identify time periods when certain activities will be done. Their strengths may be utilized in developing schedules whose exactness and precision allows workers to function in a cooperative manner. It is efficient for getting things done and it dominates people in most parts of the US and other industrialized countries.
• *Polychronism* is the perception of time as merely a context in which we live. Tasks are handled as they come. Such jobs require that the individuals constantly adjust to incoming new jobs and responsibilities. They enjoy change as part of their job (and) changing from one activity to another is the part of this pattern. This is a relationship-oriented way of living. It's efficient for building communities, personal and social relationships, and it dominates people in Asia, as well in many rural areas.

It becomes obvious that individuals who prefer a monochronistic attitude towards time will tend to feel more comfortable with traditional project planning, while polychronistic individuals will embrace agile development practice much as a matter of course.

Many of these pains can be eased when leaders clearly articulate directions and expectations in the transparent and consistent way we have described as management by objectives.

Furthermore, it can create a radical boost of motivation and a new level of collaboration once the teams understand that they are now in the driver's seat!

Last but not least, one more challenge when working within an agile project is to deal with shift of responsibility down to the development teams. A larger degree of autonomy will be especially odd if you are used to a strictly hierarchical and regulated organization. There will be a lot of uncertainty on what management is expecting and how much decision-making by the teams will be tolerated.

It should not be underestimated that management may also hesitate to embrace agile development practices wholeheartedly. The fear of loosing power and influence, latching onto familiar structures, and the pursuit of personal career goals are counterproductive to the climate of trust and collaboration which we are evangelizing.

Probably the most critical point of all is the scary feeling of surrendering a significant portion of their authority to the teams. It requires a significant amount of trust to let go of control and accept that a self-managing collaborative group requires less supervision. Let us make this clear: Rigid management hierarchies that resist experiments will prevent any attempt at applying an agile development style.

5.8 Skills

Working within a cross-disciplinary team beyond the boundaries of a specific work area will also increase the need for learning. We have already seen that soft skills and social competency are essential as soon as we lower the center of gravity in an organization and ask for collaboration and participation.

In addition, the breadth of technical expertise becomes more important. You will be confronted daily with massive amounts of information from various sources, while pursuing many tasks in parallel, using a variety of different tools. Today, a single newspaper contains more information than our ancestors of the eighteenth century would be faced with during a lifetime. New technologies will continually be created and old ones will vanish over time. The knowledge available is sky rocketing und far exceeds imagination. It is estimated that 1.5 exabytes (1.5×10^{18}) of unique information will be generated this year.

Today, these masses of information update, on a daily basis, what is state of the art. They invalidate what we learned yesterday. And they influence what we need to know and do tomorrow. Universities are preparing students for jobs that don't yet exist, using technologies that haven't yet been invented, in order to solve problems, we don't even know about yet. Companies are investing fortunes to keep up with the tremendous pace that is being set. For example, IBM spends approximately 5 billion dollars in research and development each year. This is about the budget which the Swiss government spent on education and research in 2008.

While all these aspects influence our work day in any type of IT project, especially agile software development will push the demands a little further, as agile teams claim to be self-contained units and elaborate end-to-end customer value. There is no niche limiting the personal focus on a single area of technology. Each one's technical skills need to be complemented with a broader understanding of the corporate strategy, the team's goals, the surrounding constraints, the relevant customer use cases, and a rough orientation on what is going on in the industry.

For everyone, it is definitely helpful to change their focus areas and their team association once in a while to broaden the personal experience and contribute to mixing skills within the company.

Even when accepting continuous learning as an essential part of continuous improvement, nobody will be able to pursue all facets of the activities going on in their project. Nevertheless, the ability to see the bigger picture, even if it is seen with many gaps, is an important competency. Finding a common language to collaborate and complement each other's knowledge is what makes up effective teaming.

This is true for all layers in the hierarchy. Managers will face the challenge of detailing goals with teams based on their strategic view, but without the full technical depth and insight. While in traditional handcrafts the master may have taken on the most difficult jobs himself, in today's IT industry the senior architect will delegate to subject matter experts with higher knowledge in specific areas of technology.

Further Readings

1. Hall E (1988) The silent language. Anchor Books, UK
2. Katzenbach J (2005) Wisdom of teams – creating the high performance organization. McGraw-Hill, New York, NY
3. Lewes GH (1875) Problems of Life and Mind (First Series), vol 2. Trübner, London
4. Petroski H (1992) The evolution of useful things: how everyday artifacts – from forks and pins to paper clips and zippers – came to be as they are. First Vintage Books, New York
5. Pór G. Blog of Collective Intelligence. http://www.community-intelligence.com/blogs/public/
6. Taylor FW (1964) Scientific management – comprising shop management, the principles of scientific management and testimony before the Special House Committee. Harper and Row, New York

Chapter 6
Considerations on Planning and Architecture

6.1 About Requirements, Resources, and Dates

In agile planning you are counting on what you know you definitely have. Depending on the type of project, you are either given a number of resources by your management to do the project, or, as in a services contract, there is a certain amount assigned for resources that you better not overspend. The time constraints will be fixed as well. On the other hand, requirements will be a moving target. Therefore the plan is fluid and includes only a rough estimate of those requirements that might be achievable in a prioritized order. Content is the parameter to adjust when adapting to change. The actual deliverable will be the best that can be accomplished within the given time-frame and with the given resources. Each team will figure out how to accomplish as much work as possible.

In contrast to this, traditional waterfall project management teaches just the opposite approach: You start with gathering the list of fixed requirements, then estimate the cost and time needed to implement them. This approach ignores that in most cases, resources and dates are non-negotiable given constraints. The content delivered by the team will always be confined by the team size and available time, regardless which requirements you commit to address in your specification (Fig. 6.1).

If the requirements are fixed by management as well as the dates *and* resources, this is neither waterfall nor agile, but simply unprofessional and a setup for failure. The team will be doomed to fall into a never-ending emergency mode, which is not sustainable and will offer quality decrease as the only way out.

6.2 Agile, a License for Chaos?

Critics of agile software development are very apprehensive about agility leading to chaos and anarchy: If you were not creating a detailed schedule with expected work items and estimated efforts throughout your project, you would have nothing to

T. Stober and U. Hansmann, *Agile Software Development*,
DOI 10.1007/978-3-540-70832-2_6, © Springer-Verlag Berlin Heidelberg 2010

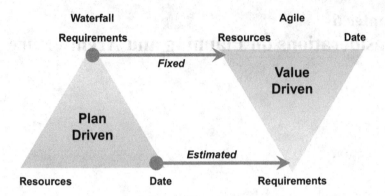

Fig. 6.1 Planning attitude in waterfall and agile projects (based on Dean Leffingwell)

measure your progress against. In an agile world, the content of a release would be a result of individuals pursuing their favorite hobbies. Without precise commitments, teams would lag behind and reduce speed, instead of adding overtime. And critics will definitely ask: "Would you really dare to travel on an airplane that was constructed following agile methodology?"

Some "supporters" might love the tempting idea that they would get away without that cumbersome documentation work. Nobody would have to worry about what might happen tomorrow. And most of all: we would not have to stand up and make any tough decisions. We would sit them out, and conveniently defer choices until issues become obsolete or decided one way or the other.

But these not uncommon views are a clear misunderstanding of the ideas behind agile software development! Do not ever ever use "agility" as an excuse for a lack of vision and strategic direction!

Let us make this perfectly clear:

Agile software development will not produce coincidental results and is not a license for chaos. Instead, agile organizations combine flexibility with structure. Processes and planning are important in order to manage and control all the loose ends that we are about to tie together.

Agility is neither "plan-driven" nor "unplanned." It is the attempt to find just the right balance. It is about *smart* planning, not about condemning structure.

The plan is useless; it's the planning that's important. (Dwight D. Eisenhower)

There is a slight difference between a "plan" and "planning" . While a plan implies something final and static, in agile software development there is much more emphasis on the aspect of continuous planning.

In an agile world, project planning is treated like everything else: it is done in a highly dynamic way, it is subject to continuous improvement, and it invites the team to actively participate. Planning covers the desire to detail, decide, commit, guide, and document whatever is needed at the time when it is needed. Just in time. Not sooner. Not later.

6.3 Balance Flexibility and Structure

6.3.1 Balance the Level of DetailError! Bookmark not defined

A traditional project plan will tell you exactly which features and tasks are planned for the entire length of the development project. The plan is typically optimized for the original goals and constraints, which were put in place at the beginning of the project. Typically a change control board will vote on any changes to the original plan that become necessary while the project proceeds.

In contrast to this, an agile planning approach will first of all differentiate between levels of detailed planning depending how far into the future you are looking. There will be a rough plan for the entire release, and a more fine-grained plan for the current iteration.

A team within an agile project can report exactly which tasks are being done in the current week, and which use cases are planned for this iteration. Beyond the current iteration, there will only be a prioritized list of use cases which are likely candidates to be covered. And everybody will understand that this list is subject to change. The further away a date is, the less detail on design and implementation will be available. When asked about the future, there will only be a clear understanding of the overall goals and general direction a team is aiming at, as well as the current prioritized list of currently known requirements.

The two-level planning of agile projects distinguishes between a high-level project plan and a fine-grained iteration plan:

- The *project plan* outlines roughly how to go from here to there, up to the point where you believe the customer will be satisfied. This includes:

  - Overall project mission, priorities, and high-level goals
  - Main focus areas and budgets
  - Assigned teams and their mission
  - Product backlog with high-level use cases
  - Initial rough effort estimates
  - Major milestones
  - Number of iterations
  - Objectives of each iteration

The most important aspect of the project plan is to provide sufficient guidance to the teams, such as which functionality is expected based on the high-level goals ("feature grid") and in which schedule the teams will execute ("timeline grid"). This kind of high-level guidance is illustrated in Fig. 6.2.

- The *iteration plan* covers only the current iteration and adds precise details who works on which tasks. This includes:

  - Team-specific iteration objectives, derived from high-level goals
  - Iteration backlog with detailed use cases

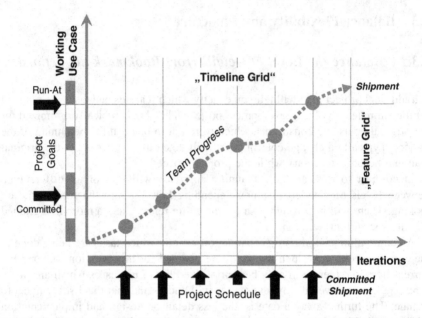

Fig. 6.2 The scope of the overall project is shaped by the "Feature grid" and "Timeline grid" in which teams can plan their iterations. Both are derived from the project mission, priorities, and high level goals

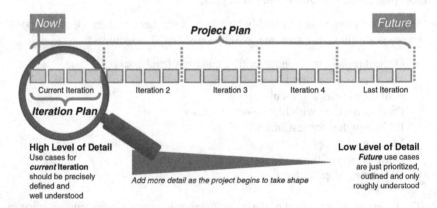

Fig. 6.3 Two-level planning of a project

- Updated effort estimates
- Evaluation and acceptance criteria.

Finding the right level of detail when looking ahead will take some experience, especially when the teams were used to a classical project management approach so far.

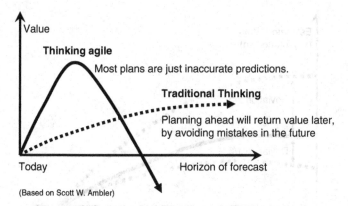

(Based on Scott W. Ambler)

Fig. 6.4 Traditional design and planning argue that you need to elaborate early and think far ahead. The value of your planning will be higher the further ahead you are able to look: since costs for fixing flaws would increase over time, your savings will be higher (dotted line). Thinking agile implies being skeptic about predicting the future, and challenges the value of looking ahead beyond the short-term future

> It is a mistake to look too far ahead. Only one link in the chain of destiny can be handled at a time. (Winston Churchill)

Try a "just-in-time" approach: At each given point in time, plan as much as is necessary with a level of precision that makes sense. Don't bother to elaborate on vague assumptions today if you will understand them much better once the product begins to take shape. Don't waste time on planning exercises in advance when it is doubtful that they will still be applicable once their time comes (Fig. 6.3).

Figs. 6.4 and 6.5 illustrate the value of planning ahead, given that assumptions about the future are very likely based on a rough guesses.

6.3.2 Balance the Timing of Decisions

Do not make debatable decisions today if they are not essential now and if there is a good chance that these decisions can be made later based on more reliable facts.

> Nothing is more terrible than activity without insight. (Thomas Carlyle)

The balance of how much of the future needs to be decided in which granularity and how many questions can be left unanswered for the time being will be achieved based on the demands of each project.

It is helpful to always keep in mind what benefit or value you would get if you made this decision right now. Or, on the other hand, what would be the cost if you left a decision pending for the time being? Indeed, deferring decisions can give you more flexibility at a first glance. But be careful: this flexibility will not always come for free!

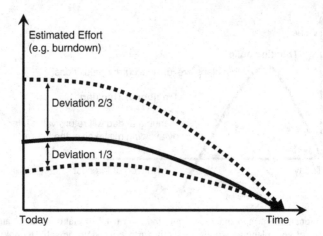

Fig. 6.5 Initial planning predictions (e.g., burndown charts or sizing estimates) will typically vary from 2/3 too high to 1/3 too small

A metaphor can nicely illustrate conflicting goals: If you are planning your summer vacation, you might want to have the maximum flexibility and leave the departure date of your flight undecided. You can easily change your reservation whenever needed, for example, you might have to fix an unexpected issue in your agile software development project. However, you will have to pay the full airfare. If you would decide on the travel dates today and waive the option of rebooking, you would be able to purchase your ticket at a highly discounted rate.

Similar examples can be found in IT projects, such as when making basic assumptions on underlying technologies or components, choosing code streams to use, or setting key architectural turnouts.

There are decisions that need to be made early, for example:

- Define strategy and high-level goals ("feature grid").
- Define iteration schedule ("timeline grid").
- Get teams staffed quickly to start executing.

Other decisions can be left pending until more information is available to decide them based on reliable facts, such as

- Which use cases from the product backlog will actually be implemented.

6.3.3 Balance the Need for Commitment

We have emphasized that organizations are focusing on vitality by optimizing their ability to quickly adapt to changing constraints. To maintain sufficient flexibility for adapting goals and plans, it is also extremely important to be very careful with commitments. Commitments will always be necessary to get the buy-in and

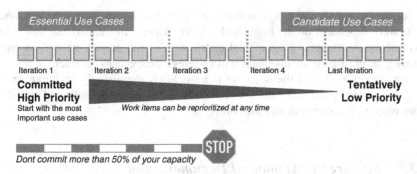

Fig. 6.6 Committed deliverables will typically be prioritized for the first iterations, while the last iteration can address stretch goals

support from the project's sponsor. But, if the given commitments imply that each conceivable hour of your team's development capacity will be required to implement them, then there isn't any room left to adjust to other incoming changes without breaking accepted dependencies.

> A first-rate organizer is never in a hurry. He is never late. He always keeps up his sleeve a margin for the unexpected. (Arnold Bennett)

The right balance between aggressive commitments and the ability to cope with changes along the way is another aspect that each project will have to optimize to fit its needs.

Do not overcommit your team. Distinguish between committed deliverables and "run-at" items. Be realistic and focus your commitment on the foreseeable short-term future. Make sure that your project has sufficient buffer to accommodate change, while keeping a sustainable pace. A rule of thumb recommends committing up to half of the team's capacity at the most. The remaining capacity will make everybody very happy as soon as additional, last minute requirements, unexpected adjustments in sizing or unforeseeable issues pop up. And these things happen, as sure as the night follows the day.

Sudden emergencies will occur and will require immediate response and significant commitment, possibly involving the full capacity of the entire team for a limited period of time. There are times when this is appropriate. Nevertheless this should be done with caution. Some organizations constantly operate in an emergency mode. In the long run, this will not be sustainable (Fig. 6.6).

6.3.4 Balance Between Autonomy and Guidance

How autonomously may a team work? How much guidance needs to be given to a team to make its members feel comfortable with their mission and to assure the project sponsor that the team is heading into the desired direction? And how much responsibility and authority is management willing to delegate?

Again, the right balance is somewhere in the middle between the extremes, and is strongly dependent on the individual project. Aspects like experience and skill level of the developers, geographical distribution of the project, as well as outsourcing or third-party involvement will be relevant aspects to consider.

But generally spoken the outcome is most likely surprisingly encouraging if project leads and managers dare to withdraw from the depth of details and give the team room to unfold talent and passion.

6.3.5 Balance the Amount of Documentation

In the previous chapter we have seen how a crisp and clear strategy, along with well-chosen goals will define the scope and direction in which everyone is heading. And we have stressed that these need to be transparent and well understood. This implies that we need to have some amount of documentation. Even agile planning will result in a variety of documents. Vitality and documentation are no contradiction. It is more a matter of the right set of documents. Documentation is important and helpful as long as each document created added value and is managed in a way that makes sense. Here are a few examples:

- Various agreements and contracts can clearly spell out goals, responsibilities, and expected deliverables at the hand-over points between different teams.
- Backlog lists and team charters will provide a prioritized list of use cases that are to be delivered.
- Continuously updated burndown charts provide data to assess the project status and help to uncover risks and delays.
- Defect documentation keeps a record of bugs that are being resolved.
- Design documents share architectural decisions and technical details with a larger community and are an excellent source of information for those who will use the resulting implementation.
- Coding guidelines are useful to achieve a common programming model. They will help to write consistent code that considers aspects like accessibility, translation, performance, etc. Furthermore, coding guidelines will make it easier for new team members to get started and will simplify maintenance in the future.
- Specifications describe programming interfaces and are needed if other teams will re-use code provided as a common component.
- Test scenarios describe test cases to be executed.
- Checklists include a lot of know-how that was gained in the past. It absolutely makes sense to leverage them to make sure that nothing has been forgotten.
- Process descriptions and documentation might be necessary to pass potential external audits in order to obtain certain certifications.

Yet again, you will have to find the right balance for your specific environment. Choosing the right amount of documentation and a reasonable level of precision will require some experimentation until a reasonable approach is found:

- Some documents might be prepared only for a certain point in time and will be outdated shortly after. Other documents might be kept up-to-date throughout the entire project. They will reflect progress and track changes of direction.
- Documents can be internal to a team, or they can be stored and archived as evidence required by any kind of certification or audit.
- Documents should be as "lean" as our organization. Any redundancy of content would be a waste of time and even a more significant waste to keep them consistent. Documents need to focus on providing value to the reader, rather than being a nuisance or part of a bureaucracy.
- Compliance requirements (audits, certification) are a driver for enhanced documentation needs.
- Sign-offs and approvals should be kept to a bare minimum. Instead, the focus should be more on collaborative creation and sharing of content.
- Suitable tooling will make document management more efficient. Conceivable tools are common team repositories, or specific project management tools, like Jazz. Wikis have turned out to be extremely efficient, as there is a minimal need for administration and it is very easy for a team to jointly add content.
- In most cases, a prototype might be a better starting point for an implementation than hundreds of pages of design. Animated Flash movies showing the user interface are an excellent way to get early feedback from the end users.

Regardless of the amount of documentation: Success is solely measured based on working and tested code. Working code is always a better proof-point of achieved results than any written status report or elaborate design document.

6.3.6 Disciplined Agile

The good news is that there is already a name for intelligently dealing with agile flexibility and predictable structure: "Disciplined Agile."

The bad news is that there is no simple and universal rule to follow, and each project will need to find out the right balance based on its needs and constraints. It will depend on the corporate culture, the team's mentality and the characteristics of the project. Mission-critical software for an airplane's navigation system will have different needs than a computer game. Typically, companies with a stronger emphasis on hierarchies will ask for a higher level of planning and documentation.

6.4 Reducing Complexity

Complexity makes it harder to understand and coordinate a project. A complex system of people, technology, and constraints is easier to control as soon as you manage to split it in smaller, digestible chunks.

Simplicity is nature's first step, and the last of art. (Philip James Bailey)

The following sections outline several aspects in which simplification will help to get closer to the vision of agile software development.

6.4.1   Simplify Prioritization and Planning

The leadership team will decide in which focus areas dedicated teams will begin to work. A very effective way to influence the investment into a product is to start defining a budget for each focus area, along with the prioritizing goals.

This high-level view on just budget and priorities will help management to outline the clear direction, but will avoid getting lost in detail. Each team will have the chance to shape the details of their focus area on its own, based on the insights and expertise they contribute to the bigger picture. This simplifies the process of elaborating the deliverables: rather then discussing all items in detail on a project-wide level, this is a more decentralized, distributed approach, which will scale much better.

Most of all, a budget will translate into the size of a team which is in charge of a particular focus area. The larger a team is, the more content will be delivered in that particular focus area. In an agile environment it is obvious that the budgets and the prioritized product backlog will adjust whenever necessary. For instance, slow progress in a strategic focus area might require the rebalancing of resources to strengthen the team driving that area. Of course, this rebalancing will imply that other teams will be able to complete less use cases than they had anticipated.

As far as the planning of the execution is concerned, the project will be a sequence of short and time-boxed iterations. The iterations will compose the common rhythm of all teams and resemble the heartbeat of the project. Each iteration will deliver use cases, which have been completed and tested by the teams and will add value to the product. One technical challenge is to cut large work items into digestible pieces fitting into a single iteration. The smaller the chunks, the easier it will be to manage the work and track progress and to understand design, tests, and dependencies. The bigger the identified tasks are, the more difficult and imprecise will be their sizing estimates. A rule of thumb is to identify use cases and divide them into smaller tasks that can be designed and completed within a few days or 1 week at the most.

"Divide your movements into easy-to-do sections. If you fail, divide again." (Peter Zarlenga, Founder of Blockbuster Videos)

"The secret of getting ahead is getting started. The secret of getting started is breaking your complex overwhelming tasks into small manageable tasks, and then starting on the first one." (Mark Twain)

It is easy to monitor each task as a single unit if the chunks of work are kept as small as possible and their status is either "not started," "in progress," or "completed." For the purpose of planning and project management, we can view a small task as a very fine-grained "black box," which we can juggle around extremely flexibly. As soon as tasks become less granular and resemble more work, it becomes necessary to look into the status in a much more detailed way. For instance, a status report would need to guess to which amount the work is completed and what is left to be done. Such estimates are error-prone and time-consuming.

While planning in general will always keep an eye on the priorities and budgets which have been set, the literal project management will focus only on the current iteration.

6.4.2 Simplify Team Setup and Dependencies

As part of your planning activities, you need to structure your organization in a manageable and simple way. We have discussed this earlier in detail: Agile software development suggests establishing teams, which work in an interdisciplinary fashion and focus on implementing their use cases end-to-end. Once goals are agreed upon, teams should be able to pursue their own endeavors in as focused and independently as possible. In an ideal world a team would get started once the suppliers make all necessary input available, and the team would be done as soon as it had delivered its final output to the customer.

This is, of course, not realistic or pragmatic. In reality, some overlap between two teams can hardly be avoided. Teams will work in parallel on the same common components. Despite the fact that they will pursue use cases end-to-end, there will still be functionality that they will consume from others, or that they will provide to others. But driving such dependencies directly between the involved customer and supplier will be simpler and more efficient than coordinating the dependencies within the entire project with a central project management.

How can we support more autonomy and minimize the dependencies?

The first approach is partitioning the teams within a project intelligently: You can reflect geographic constraints and co-locate those developers assigned to the same team. You can also reflect the existing skills and experience when assigning focus areas and use cases to teams. But always keep in mind that for every nontrivial project, it is impossible to partition the work cleanly and there will be compromises and sacrifices.

The second approach is to reduce dependencies between teams as much as possible. Minimizing dependencies will simplify the efforts spent on agreeing on interfaces, communicating hand-over, and coordinating the timing for the exchange of deliverables. And it will prevent blocking the progress of one team when a required prerequisite is not delivered by another team on time.

Coordination of a project will become more difficult the more prerequisites need to be spelled out.

There will be functional dependencies and scheduling dependencies.

Functional dependencies arise from the work split when two teams need to modify the same area of the code base or need to come up with a common definition of a service or interface. These dependencies need to be addressed by the software architect. A service-oriented architecture, componentization efforts, or a plug-in concept will help to decouple code and allow for a more independent development of components. It may also be possible to choose the focus of different teams in a way that they have no or little overlap in terms of code access. We will focus on architectural considerations later in this chapter.

Scheduling dependencies will occur once one team needs to complete certain functionality before the other team can pick it up and proceed. If such dependencies cannot be avoided, it is advisable to have single iterations as the smallest unit of time in the schedule planning and bear in mind that the work beyond the current iteration might change in content or timing. It is also very helpful to have a staggered start of the project: teams that work on very fundamental capabilities such as install or platform infrastructure can begin earlier and thus provide a more stable baseline once the other teams join in. The same applies to very disruptive changes.

Any dependencies that are unavoidable need to be closely monitored by the teams involved. They will make up the critical path of the project.

6.4.3 Simplify Tools and Processes

In general, processes make a lot of sense. Customer relationship management or managing the life-cycle and versioning of code will not be possible without.

However, there is the risk that every process tries to be as generic and complete as possible. As a consequence, it will grow in complexity. In a "one-size-fits-all" process, there will be many exceptions and special cases to deal with. Over time, people won't be able to follow the process, as its description is too complicated or cumbersome. Eventually, the process will be out of sync with what the team is actually doing. The gap between project and process will become wider and wider, until the process dies under its own weight.

To enhance agility, it is advisable to revisit the existing set of processes and streamline them: Which ones do not provide enough value? Which ones are too cumbersome and distract more people than necessary? Which ones are disconnected from the actual development reality?

For instance, some kind of software design process will be required to ensure that an architecture document is communicated to all those who will be impacted and need to sign off. Would a shorter list of reviewers accelerate processing and turnaround time without adding too much risk of oversights? Is a Wiki a better way to jointly elaborate design documents than PowerPoint files? Which types of development work require a formal design and which do not?

Another important process within project management is the tracking of progress: As we are going forward with continuous planning, we need to be vigilant and monitor the performance of the teams. A common weakness in software development is the identification of parameters to assess performance of teams and individuals. While a classical manufacturing plant can easily measure throughput and costs per produced unit, it is not that easy in the IT industry. Helpful metrics for this purpose are defect statistics or burndown charts. Any of this data will reflect the knowledge we have today based on judgments and guesswork. And this data will be updated constantly and provide more precision as the project proceeds and the remaining uncertainties begin to go away. The tracking of progress is a complex and important responsibility. But it must not necessarily end up in a cumbersome process nightmare. Scrum is an excellent example of how process management can be done in a very efficient, reliable, and extremely lightweight way.

Tools will help to automate processes, manage work items, compile code, execute test cases, count bugs, and also make documentation and tracking much easier. But whenever you're dealing with tools and processes, keep in mind that it always needs to be as easy as possible to continuously update information at any time: By design, changing a plan is part of living in a nondeterministic world and is part of our daily work. It is not a sign of failure.

But no matter which processes are actually implemented within a project: In the end, it is only working software that really counts! And remember – agile software development clearly states that interactions and individuals are valued above processes and tools.

In many cases, the team's way of working is based on a grown mix of best practices, which are driven by practitioners and are not reflected in any formal and generic processes at all. "Grow and evolve" and "mix and match" are two very fitting slogans to think of when setting up development processes that are to meet the needs of an individual project and provide real value.

6.4.4 Simplify Daily Life

One risk of decentralizing project management is that we will grow a confusing dark jungle of miscellaneous tools, establish inconsistent processes, and live differing teaming cultures. This will not scale once individual teams begin to deployg and administer their own choice of tools and define their own processes.

An established common infrastructure of services will help make daily life easier for everyone. Such an infrastructure will unburden the teams from redundantly creating their own setup and it will enforce a consistent approach across the project or company. The most relevant services are the following:

- Connect people in a decentralized world is crucial. Obviously, powerful communication and collaboration platforms, with document repositories, Wikis, and instant messaging are extremely valuable. This is especially true for teams that are not co-located.
- Providing a build environment as a centralized infrastructure makes sense as well. This includes a code repository with versioning capabilities, the compilation of code, and the execution of automated regression tests.

- Using a common defect tracking tool is highly recommended. It helps keeping track of defects and gives an indication of the quality of the code delivered.
- Establishing a common development tooling environment will boost efficiency. This aspect includes the development tool itself, but also functionality such as quick download of the latest code base and test code changes.
- Setting up a test system with the most recent build is a daily job to do on many machines. Especially in complex environments with a stack of software products, this can be extremely time-consuming. Automation will be crucial here.
- Exposing a transparent and up-to-date project status to all team members almost goes without saying. Project management features, like listing use cases and work items and tracking their progress, can be made available to everyone by a common set of tooling.

Such common services are not only tooling and IT services. They can also include certain highly specialized skills that cannot be provided by each and every development team itself. One such skill is a performance expert who consults teams and resolves performance-specific problems.

Although teams own the responsibility to deliver working and tested code, it may make sense to establish additional centralized testing services. System verification testing extends the test coverage of the teams and verifies that the software will run on all intended platforms as well as in complex environments such as server clusters.

Last but not least, existing information isn't monopolized. Instead, it needs to be made available to the entire project for better transparency. Secrecy within a decentralized project is not helpful. This applies to design as well as project planning-related information.

6.5   Architectural Considerations

6.5.1   Outside-In Design

Focusing on the needs of a customer is the key to a successful project. A software solution must not only be free of bugs, it must also match the customer's expectations. But how can we gather customer requirements and ensure that they are translated into the right set of use cases to be delivered?

Outside-In Design is a set of methodologies for bringing the needs of customers into the development process.

We need to understand the context in which our products are used to make them better, more sellable, and usable. Customer requirements are the beginning of any development activity. This includes understanding the needs of end users. This can happen with the help of interviews, discussions, surveys, or by experiences gathered in various kinds of customer-facing activities.

Early customer feedback is the second major element in Outside-In Design. The range of options includes sharing of design documents or user interface sketches, as well as beta drops and pilot installations. Feedback can be qualitative or quantitative. It is also possible to actively involve customers in the development teams. "On-site customer" is a technique that is part of Extreme Programming.

Feedback can help improve a product and its capabilities, but it can also point the development team to new ideas on how modify and extend existing products in new and useful ways.

Once again, IT infrastructure featuring Web 2.0 applications, like a Wiki, will open up new ways to let external customers easily participate in design work and open discussions.

6.5.2 Requirements, Use Cases, Scenarios, and Tasks

Based on a customer-driven view on everything, content to be delivered by a product or project is described from a user perspective that outlines how the deliverable needs to behave in order to meet the customers' expectations.

An essential artifact for this purpose is a use case. Use cases detail functionality from the user's point of view and clarify "what" will be done. Sometimes the term user story is used in lieu of a use case. A use case is the common and easily understandable language when discussing deliverables and content between customers, architects, developers, testers, and project management. Each use case can address one or more formal requirements, which may be listed in a specification of work.

A use case description lists the set of interactions between the developed system and the involved actors. It answers questions like "Which results must a system provide?" "How will these results be triggered by the users?" The use case description can summarize this in a simple way by stating involved role(s), achieved goal(s), and delivered business value. For instance, an e-travel application might include the following use case:

• A customer (= role) can pay for his flight with a coupon (= goal) to immediately benefit from the company's loyalty program (= value).

Use cases can be documented and detailed in plain text. Modeling tools help describe use cases in a more formal way, for example by leveraging UML diagrams.

Use cases can be described at different levels of detail. They typically start very lightweight as a rough idea. This is the right precision during the high-level planning the overall release, when you establish the product backlog. Over time, use cases will evolve by adding more and more detail, as the use case and its context and dependencies become clearer and better understood.

In our sample use case above, we might need to differentiate between a registered user, a gold-customer, and an anonymous guest. Or there might be different kinds of coupons.

Use cases should be rather small. They can be implemented and delivered preferably in less than a week, but at least within a single iteration. If complexity grows, it is time to split larger use cases into smaller ones. The use case "Book a travel" is far too complex. Here are some suggestions how to break down such an epic use case into smaller chunks which will be easier to manage:

- Look for finer grained user roles, such as a "gold-customer" or a "guest user," which slightly differ from each other. Define a specific use case for each role
- Find boundaries of the used data, such as reservation or payment data. Define specific use cases for each kind of data or system you are accessing.
- Find boundaries of the organization, such as customer support, procurement, or technical help-desk. Define a separate use case for the interaction with each organizational unit.
- Stage different aspects or features of a use case and deal with each separately one after each other. For instance the ability to make a seat reservation can be added at a later iteration.

When detailing use cases, it is helpful to identify a set of different scenarios for each use case. Each scenario covers one specific walkthrough and path through the use case.

Let us have a look at our sample use case once more. We could envision the following scenarios:

- The coupon has expired and the user will receive an error message.
- The value of the coupon is not sufficient to cover the bill. The user will need to enter a credit card to cover the remaining amount.
- The coupon number is invalid. The user will need to reenter the coupon number.

The use case and its scenarios will be fully elaborated within the iteration in which the use case is implemented.

When defining a use case and its scenario, it is also helpful to define acceptance criteria for its implementation that validate whether or not a system works correctly. Acceptance criteria can include a set of input values and the expected results. They can also include test cases to address the nonfunctional as well as the functional requirements. Test cases are an integral part of a use case! It is advisable to specify acceptance criteria and test cases before implementation starts. They will make it possible to immediately test any code drop delivered.

A use case description does contain some ambiguity on "how" this will be translated into technical terms and what the implementation will look like.

Therefore, each use case is broken down into smaller, separately deliverable tasks that describe the work items required to implement it.

This level of detail is only required for the work selected for the current iteration. The implementation effort for each individual task can be sized fairly accurately, since by now the context of the use case is well understood and the complexity of a single task is rather moderate. Tasks can be assigned to individual team members. They should be rather small chunks of work and be implementable in a few days.

6.5.3 Architectural Attitude

Creating customer value means implementing working use cases that address the needs and requirements of end-users. While the software architecture itself is, generally spoken, irrelevant to the end-user, it is extremely important for most other stakeholders: designers, developers, and testers will benefit from understanding the architectural approach. They will write code and test cases which follow defined guidelines and which fit into the framework of the underlying architecture. Architecture and development strongly interact with each other. New implementations suggest architectural changes. Architectural changes usually require radical changes to the implementation.

If a system's architecture is entirely accidental, this might be acceptable for small projects with a short lifetime of the resulting deliverable. Larger systems will need an intentional architecture that is thought through and will scale with the future needs while it evolves.

Just as any kind of planning and design required documentation, so does architecture. Again, agile development can be spontaneous, but is not an excuse for working sloppily. A disciplined agile project will produce high-level architecture overviews, use case descriptions, contractual documents, and API specifications. The key is simply to make architecture and its documentation as lean and painless as possible. Like project planning, the architecture and design work is a continuously ongoing task.

Well-architected systems ought to be simple to describe. They can be extended as needed, and are built from components that can be implemented independently. Within agile development projects, architectural considerations are similar to those proposed for all other kinds of planning activities:

- *Keep it simple:*
  A complex system should be split into smaller subsystems that allow creative teams to work concurrently and independently. While there needs to be an overarching architectural concept such as a service-oriented architecture or a set of layers that make up the software stack, there will be an independent architecture and design effort for each individual component. The goal is to create componentized software pieces or services that can be better understood, implemented more easily and managed more flexibly. The more functional dependencies there are between components, the more difficult it will be to design enhancements and to coordinate the implementation work. In addition, more planning will be needed to align the different teams with each other.The architecture of the software solution should be simple, easy to understand, and as straightforward as possible. The same is true for coding: The code should be kept clear and simple. Keeping it simple also means avoiding unnecessary generalization. Features that do not address customer needs and are not derived from customer requirements will add to development, test, and most of all maintenance costs without adding significant value. In terms of lean software development, this is avoidable "waste." Great software that can be customized with a

vast array of parameters may be adaptable to many situations, but it will likely be complex to set up for a customer, and will be tedious to maintain.

- *Keep it flexible:*
  Then again, a versatile product needs to be highly customizable to fit into many different environments. In addition, a good architecture needs to be flexible enough to accommodate future extensions, plan changes, and new requirements. You need to keep in mind that new technologies evolve, some customer priorities will shift, and additional integration demands with other products will arise. Today's favorite concepts may have to be changed in the future. Yet again, we need to find the right balance between a simple and straightforward architecture on the one hand and an extensible, generic platform on the other hand.

- *Keep it small:*
  Any kind of architecture and design work needs to be split into smaller pieces that can be described and implemented within a single iteration. A system grows by incrementally adding small executable and testable use cases. Making big architectural changes in small and safe steps is one step toward excellence in agile software development. It takes a lot of experience to find suitable options for dividing large changes into digestible chunks. This is especially challenging at the beginning of a project, when the underlying infrastructure, like install procedures, or support for a new operating system is established. It may be difficult to show any preliminary results before the majority of the work is completed. Scaffolding code can be useful here to run or demonstrate an intermediate version of the product, despite the fact that crucial parts are still missing.

- *Keep it transparent:*
  Obviously, good communication is key to getting the buy-in for an architectural approach: A design needs to be discussed in time with the people who need to understand it. It needs to be documented. A diversified set of domain experts who bring the right set of skills to the table can come up with a single, coherent architecture. They might have feedback or will need the proper explanations to get started. This process of socializing a design will also be a driving force for cross-team integration and is an excellent way to identify functional dependencies between teams.

- *Keep it vital:*
  It is a myth that architecture is something that must be complete before any development takes place. Actually, there will be significant overlap with the resulting implementation work. Instead of investing months in building a detailed specification, the teams will continuously add to a baseline. Architecture and design will never be done, but will be continuously refined and improved as the product evolves.

Similar to the agile project planning approach based on goals and strategies, the architecture will primarily outline a rough direction, while the actual design details of various aspects and components will shape out as the product evolves. Again, the key to success is finding a suitable balance between timely decisions and leaving

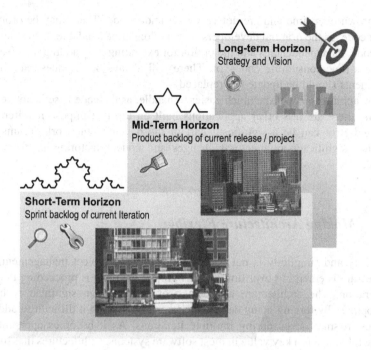

Fig. 6.7 Planning detail is differentiated, depending on the time horizon

options pending, between directive guidance by technical leaders and team autonomy, and between comprehensive design documentation and hands-on prototyping. Yet again, there are no general rules.

Many things difficult in design prove easy in performance. (Samuel Johnson)

Agility also implies that the level of detail in which architecture is understood and described will depend on the period of time you are looking ahead. The further you look into the future, the vaguer a design will be, the more alternatives will be left undecided, and the more unclear will the details of the actual implementation be. While fundamental design assumptions, like the technology platform or the code stream to be used for development, need to be decided early, other choices like the layout of the user interface can be postponed.

Same as the project management is done on two levels of granularity, architecture and design work will also need to find the right balance of detail and planning ahead (Fig. 6.7).

Architectural demands will strongly differ from project to project.

Mature systems typically have quite a stable architecture. They have grown over long periods of time and represent a proven solution on which many customers rely. It will be challenging to adapt and change such a system without breaking legacy APIs or usage patterns that have been heavily applied by customers. A code base that has evolved over many releases will likely have difficult dependencies that have grown over time and are not very well understood. There may be orphaned source code, as the original developers may not longer be available. Understanding such code, its dependencies and the option for extending it by adding new features will be a time-consuming endeavor. There will always be the question whether components should be updated or replaced.

New applications will be much easier to handle, as the code base is not yet that big, and there are less other applications built on top that impose requirements. Trial-and-error can be a significant part of the development work. Teams may introduce significant architectural changes and code refactoring in every new release.

6.5.4   Making Architecture Flexible

Flexibility and simplicity is not just a matter of agile project management. The architecture is emerging over time, just as project planning is proceeding iteration by iteration. The architecture and design of software can significantly hinder development by forcing many dependencies and by making it difficult to add new features in small steps during multiple iterations. As it becomes apparent that change tolerance is a key value in most software systems, architectures that support incremental development are clearly superior to monolithic ones.

> Complex systems have somehow acquired the ability to bring order and chaos into a special kind of balance. (Mitchell Waldrop)

One example of a flexible pattern is a service-oriented architecture (SOA). SOA suggests building business logic from reusable services, such as repeatable tasks. A high degree of reuse can be achieved by linking services and their outcome to composite applications. SOA includes patterns defining how a service needs to behave and how to connect services with each other. The most mature architecture is the concept of an enterprise service bus, which is a skeleton into which the services can be assembled and integrated in an extensible way. The Enterprise Service Bus provides messaging and communication capabilities within a software system, as well as mediation and transformation between the available services. Other examples of service-oriented technologies are plug-in architectures that are used, for instance, in browsers or the extension registry used by Eclipse.

Fig. 6.8 shows an example of a service oriented architecture.

A second pattern is that of componentization. Componentization is a best practice that any new development project needs to consider. The goal is to provide flexible independent components that can be reused to a large extent. Each component exposes a well-defined contract defining how it must be used. Different applications can be assembled from an infrastructure of common components.

Componentization is also relevant for mature systems: existing monolithic software products can be decomposed into distinct components that can be reassembled into new offerings (Fig. 6.9).

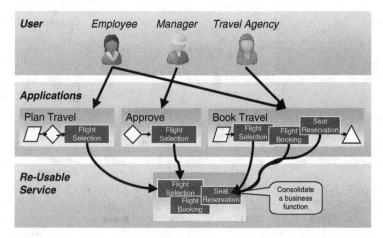

Fig. 6.8 Service oriented architecture (SOA): the end-user facing application is assembled from a set of reusable services which are combined

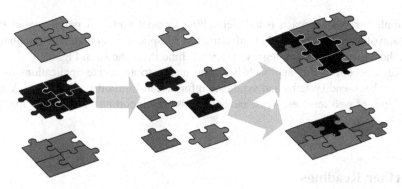

Fig. 6.9 Software componentization: decompose existing software into reusable building bricks, which can be assembled to new offerings

Rather than creating one huge product that is highly customizable to meet all needs ("one size fits all"), componentization aims at identifying a set of fine-grained and flexible building bricks from which customer -specific solutions can be assembled rather quickly. The architectural focus is on the assembly of applications from reusable assets, not on understanding the implementation details of the individual components.

Tricky aspects with any componentization effort are common deployment, installation, versioning, and dependency management. It is important to keep individual components as independent as possible and to avoid build as well as runtime dependencies.

The third pattern we want to outline is the pattern of integrating inhomogeneous subsystems. In this case the existing subsystems will not be changed, but instead a new system will be added to act as an integrated front-end to the end-user. A good

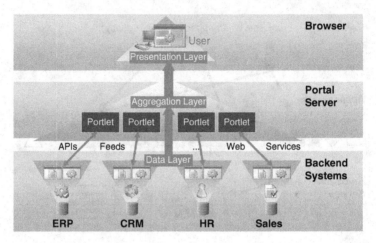

Fig. 6.10 A Portal integrates application and information from various sources and gives the illusion of a nicely integrated system

example for this approach is an overarching Intranet Portal: A portal creates the illusion of a nicely integrated IT infrastructure, despite the fact that there is actually an inhomogeneous mix of legacy systems behind it, as shown in Fig. 6.10.

Key concept of this approach is to find means to aggregate applications from various back-end systems and syndicate information from different sources. Various kinds of web services, APIs, and data feeds are supporting technologies applied in this context.

Further Readings

1. Ambler S (2002) Agile modeling – effective practices for extreme programming and the unified process. Wiley, New York
2. Cohn M (2005) Agile estimating and planning. Prentice Hall, Englewood Cliffs, NJ
3. Coplien J, Harrison N (2004) Organizational patterns of agile software development. Prentice Hall, Englewood Cliffs, NJ
4. Leffingwell D (2007) Scaling software agility: best practices for large enterprises. Addison-Wesley Longman, Amsterdam
5. Sweitzer J (2007) Outside-in software development: a practical approach to building successful stakeholder-based products. IBM Press, USA

Chapter 7
Considerations on Project Execution

7.1 The Big Bang

Friday, September 28, 2007, was a sunny, warm day in autumn. The leaves were starting to turn from a luscious green to shades of yellow and bright red. That Friday had been defined as the milestone that completed the precisely scheduled development phase of a nonagile, stage-gated project. On this milestone, programmers were supposed to hand over their achievements to the test team. The project managers had been reporting a reassuring green status in their reports over the past weeks. Their good-looking PowerPoint charts had been passed up the management chain and inspired great confidence. The entire team was looking forward to a relaxing weekend, as the last weeks of the development phase had been extremely strenuous. In long hours, everyone had been rushing to complete the work that had been assigned to them. The developers had managed to check-in an incredible amount of last-minute code changes into the source control system.

But there was one ugly detail: Many lines of code had been accumulated off-line and were now added to the main code stream just in time to meet the deadline of the milestone. The build stability decreased rapidly. Build breaks occurred more often, and even once the basic compiler errors were resolved, the product was still not installable. Just too much code had been added concurrently. Interfaces of components were out sync, and version conflicts had to be resolved. And worst of all: there had not been enough focus on testing. Especially the integration testing across the functional units had been gravely neglected.

But in spite of these problems, on that particular Friday, the coding was complete, as scheduled and promised.

In the week after the milestone, the build team and the lead developers started to work through the issues and resolved one problem, just to find the next one. What followed was a veritable nightmare: More and more experts were added to a task force, which struggled for many weeks to get at least an initial version on the product up and running. The task force surfaced that a major redesign in the product's architecture hadn't been reflected by some components due to lack of

T. Stober and U. Hansmann, *Agile Software Development*,
DOI 10.1007/978-3-540-70832-2_7, © Springer-Verlag Berlin Heidelberg 2010

communication. This disconnect could not have been discovered earlier as teams were thinking and acting within their constrained functional boundaries, and without thinking in the corresponding use cases end-to-end. There have been no iterations of working code during the development phase that could have been used to validate design assumptions and test the deliverables already in early stages.

During this time, the leaves in the forest, just like the status of the project, turned into a dark and dangerous red. It is not that the project management had not foreseen the risk of some integration problems – they had scheduled 2 full weeks for an integration phase. But the integration took much longer than anyone had anticipated. The task force was still working through the issues when outside all leaves had fallen.

7.2 Continuous Integration

This example may appear to be fairly extreme, but it is not at all unusual. Prioritizing the quality of working code higher than adding new functionality may be a challenge for many developers who are ambitious and eager to push cool and creative new features rather than dealing with tedious testing.

But the critical situation at the end of the development phase can be avoided.

One part of the answer is to test each single code change thoroughly and immediately. If comprehensive regression tests are automated, they can be executed quickly and efficiently throughout the iterations. Ideally, these tests are part of the build process. Testing includes unit tests, but also more complex functional verification testing. Even early stress, load, or performance testing can help to address stability issues or performance degradation early in the game. Test results will provide a reliable and detailed quality status. They can alert the teams if the quality gets worse.

The second part of the answer is to make sure that detected bugs are fixed rather quickly, instead of accumulating them to a significant backlog of issues that cannot be handled any more because we are running out of time. Resolving pending quality issues has a higher priority than proceeding with new functionality.

And finally, one important aid in maintaining stability is to add new functionality incrementally in small chunks. Development proceeds by designing and implementing small chunks of code, which are immediately tested and continuously integrated into the code base, without lag time in between.

In this context, the usage of source control and build system play a major role: The teams will work on a single code stream. Developers need to continuously integrate and commit even small code changes to the common code stream immediately, rather than implementing on a parallel code stream and merging a batch of changes at a later time. The common code stream should always reflect the most recent and complete working version of the software. Accumulating code changes locally without integrating those increases the risk of conflicts and incompatibilities with other concurrent code changes, which are hard to resolve. They are even harder to resolve later in the cycle, when it is more difficult to determine the actual change that caused the problem.

At any time, it needs to be possible to build the entire system from a fresh checkout of the latest and complete source without any further dependencies. The build needs to be fast and executed continuously. Based on the output of each build, a set of automated regression tests need to ensure that the code changes did not cause any side effects. Finding bugs, i.e. integration issues early, makes it easier to resolve them quickly: The teams are still very well aware of the code they have touched last and they have the context of their code changes still fresh in their minds.

The compiled output also needs to be made available to testers and stakeholders immediately in order to get quick feedback on whether the working code fulfills the requirements and to verify that the quality meets expectations.

In short: continuous code integration, regular builds, and immediate feedback will help detect and fix integration problems. And most of all, they will help avoid the last-minute stress before iteration exit or release shipment.

Continuous integration will only work well if the quick turnaround and pace of the development activities can be sustained. A long execution time of builds or automated test suites, compiler errors and failing builds, and not-installable product drops will be extremely disruptive and will cause a quickly growing backlog of code changes. The backlog can then not be integrated and tested, as no current and working version of the product exists. Our experience has shown that the likelihood of build breaks increases exponentially with the number of fixes in the backlog. "Never break the build" is the principle rule to obey in this context. This is much more important than completing a development feature at a certain promised date. Indeed, breaking the build is worse than missing a date. You will get another chance to deliver your missing piece within the next iteration, whereas a broken build will delay all deliverables of the entire project. Incomplete or even disruptive changes can be hidden e.g. by disabling the new code path by default.

Continuous integration fits nicely into the agile model, in which continuous improvement is a general theme:

- Continuous iterative and adaptive planning
- Continuous design
- Continuous testing (automated)
- Continuous listening
- Continuous conversation and collaboration
- Continuous demos
- Continuous consumption of our own output
- Continuous status
- Continuous feedback
- Continuous learning
- Continuous progress

We have seen that a considerable amount of trust is needed by management to surrender control and authority directly to the agile teams. This trust will be reinforced daily with visible progress and continuously working code.

Having working code is the safest and most efficient way to surface the level of quality and to validate early design assumptions. Working, tested, and integrated code is the only unambiguous criteria for being successful. It is also extremely helpful to show and tell others what has been accomplished so far. In contrast to a PowerPoint status report, working code will keep everyone honest and present unfiltered, first-hand truth. Project tracking will be simpler, as the daily build quality will illustrate the progress.

7.3   The Rhythm of the Project: Iterations

We want to reemphasize the importance of developing in small incremental iterations. New features will be added to the project deliverables one step at a time. And each step includes design, coding, unit testing, as well as functional verification testing.

Any code change is continuously integrated throughout the iteration. While a code change needs to result in a stable overall system, each iteration will achieve a level of quality that serves as a baseline for the following iteration. Ideally the quality would even allow a customer to install and evaluate the deliverable right away.

There is always much debate on the perfect duration of an iteration. Two weeks is a fairly common rule of thumb. Obviously, the duration should be rather short, in order to avoid creating a mini-release with a huge milestone at the end. Therefore, 6 weeks is far too long, and will quickly turn into a waterfall model. If you can't get something done in one iteration, the next opportunity to complete and deliver it within the next iteration shouldn't be too far away. If you have to wait for several weeks, there is the temptation to rush through testing a bit too hastily and thus risk sacrificing quality.

The more developers are producing code, the more changes get added to the code base and increase the need for integration and stabilization. The more time allotted to integration and stabilization, the shorter iterations should be in order to avoid getting overwhelmed by the amount and complexity of new code.

Even more important than the actual duration of each iteration is the fact that iterations should be time-boxed. They have a fixed length, their start and end dates are synchronized across all teams. While the time allotted to an iteration is a given and will never change, the amount of delivered content within an iteration depends on the actual speed of progress. In other words: the length of an iteration is confined by time. The amount of delivered content is confined by the need to deliver quality code.

The idea of iterations is to create an even and steady rhythm of progress in a sustainable, constant pace – repetitively and predictably. All teams and their daily work are synchronized in the common schedule of iterations – the heartbeat of the project. You plan at the beginning. You perform and deliver throughout the iteration. At the end you will close the development work and defer any uncompleted items to the next iteration. As you produce the stable final output of the iteration, the progress of the project will pause for an instant of time, just to resume quickly with the next iteration.

Especially at the beginning and end of the iteration the pace can easily get out of control, especially if the duration of the iteration is rather long and the teams tend to commit to work beyond their capacity. If you have worked through significant amount of code changes and you start too late with wrapping up and testing, the end of an iteration will be a painful struggle to get the code integrated and working. Even if you are able to get a stable driver after several days and nights of hard work, the teams will be exhausted and will start the next iteration at a slower pace. This effect will ripple from iteration to iteration and the project will not get to that even and steady pace. Performance will drop. The pace will be jumping between extremes and the team's progress will suffer from heart arrhythmia (Fig. 7.1).

Iterations are planned in a much more fine-grained way than releases. In a two-level planning approach, the level of detail differentiates between a long-term and a short term-planning horizon. While the release plan outlines a rough schedule and the product backlog with the prioritized requirements to be addressed, the iteration plan only focuses on the requirements to be implemented in this particular iteration. The iteration is kicked-off with elaborating use cases for exactly those selected requirements.

The developers pick up use cases from the iteration backlog strictly in the order of priority. Each use case will be designed, implemented, tested, and integrated into the code base. Any occurring problems will be fixed right away to maintain a reliable and stable code throughout the iteration (Fig. 7.2).

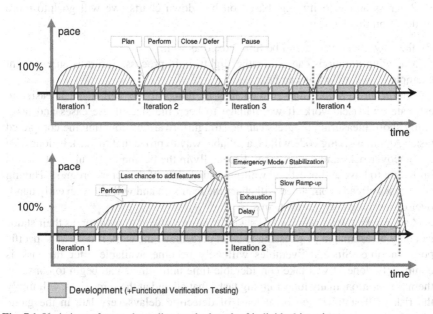

Fig. 7.1 Variations of pace, depending on the length of individual iterations

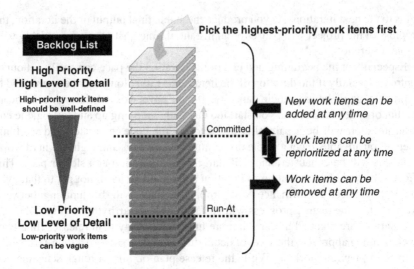

Fig. 7.2 Working through the backlog in order of priority

Progress is monitored in daily, but brief stand-up meetings. The most important data to assess the current status includes:

- Achievements, issues, and next work items (aka as the three Scrum questions)
- Today's build quality (is the code working?)
- Today's results of the automated regression tests
- Progress and velocity, e.g. based on burndown charts (we will go into more detail on this later).

By the way: meetings should be time-boxed as well.

Managing iterations can be a rather lightweight process. Scrum is an excellent example how simple this can be.

Within an iteration, fine-grained lists of things to do will make it easier to estimate and track work. If we manage to keep the size of use cases and tasks really short, measuring progress can be straightforward by counting the completed tasks. Again, working code will be a reliable way to prove that the task is done. We can uncover risks and delays early. Especially in the beginning of an iteration it is important to have an honest and well-grounded view of the actual progress. Having some tasks already completed will cheer the teams on and will allow an early hand-over of working code to test teams.

As a contrast, long running tasks will be more difficult to monitor, as their status can only be estimated by guessing to what percentage the work is done at a specific point in time. Since deliverables will only become available after the task is completely done, it will take considerable time until others can begin to consume them. In addition, many long-running tasks that all end at the same time, will imply the risk of destabilization or at least of detecting delays very late in the game (Fig. 7.3).

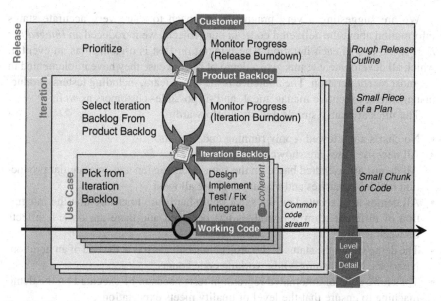

Fig. 7.3 Creating working code is a sequence of steps, detailing a rough release outline into short iterations and finally implementing small chunks of working and integrated code

Once iterations run smoothly in a steady pace and continuously produce stable code, experienced agile development teams can increase the rate of strokes by making iterations even shorter. The pain of wrapping up the project, by packaging the working code into a formal product release will become rather insignificant as well. Getting to the next level of flexibility can be to ship smaller and much more frequent releases.

7.4 Integration Fest

During the last few days of the iteration, the teams will need to focus on consolidating the code. Thanks to continuous integration and a limited amount of changes due to the short iteration length, that consolidation will be manageable.

Acceptance criteria of deliverables at the end on an iteration typically are:

- User stories are implemented and complete
- Critical defects are fixed
- All defined test cases are automated and pass successfully
- All necessary end-user documentation is written.

In short: Items that are considered as delivered really need to be completely *done* at that time.

An important result of each iteration is a shared understanding of what functionality is sound and trusted, and which area of the evolving software base is still work in progress.

We are suggesting a very pragmatic approach to share very accurate status information about the delivered code: in our projects, we introduced an *Integration Fest* at the end of each iteration. The Integration Fest is organized as an event in which all development teams give a demo of the use case they have implemented in the most recent iteration. The audience is the entire team, including testers, product management, executive management, and -if possible- customers as well.

The rules are fairly simple and straightforward:

- No charts are allowed – only running code counts.
- All new use cases are shown.
- All demos are executed based on the same code version and on a regular production build. No patches and manual fixes are allowed.
- All demos are executed on the same demo hardware to ensure that the integration of different team deliverables has occurred and there are no side effects between teams.
- The demo shows the status of the project honestly, as it is at the end of an iteration. No tricks. No polishing. Only working code counts.
- The entire suite of automated regression tests needs to be executed on the demo machine to ensure that the level of quality meets expectations.

The benefits of the Integration Fest are very valuable:

- For the developers, the Integration Fest is an excellent opportunity to regularly show what they have achieved. They get the chance to convey trust and confidence in the code quality. In addition, demonstrating working code and its value is a good source of motivation.
- This interlock is also a practical source of information for anyone who wants to learn more about the functionality and content of the product. At the same time, the Integration Fest is a skill transfer and education session and helps to better understand what the other teams are doing. Especially for independent test teams, it is extremely helpful to get started with new functionality quickly.
- The project manager gets an honest status of the project's progress based on working code.
- And finally, the integration fest is also beneficial to the relationship with a customer, who will, in regular intervals, see what progress has been made. The interlock will help to validate early if the deliverables really meet the customers' requirements.

The following figure shows the timeline of iterations and the scheduled integration fests (Fig. 7.4).

7.5 Juggling Content Within an Iteration

Once the goals of a team are agreed on with management, the teams will begin to shape requirements and use cases they intend to pursue in order to address their goals. As part of this process, they will work with product management or involve

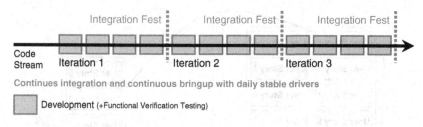

Fig. 7.4 Integration Fest

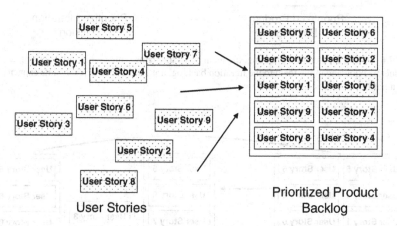

Fig. 7.5 At the beginning of a project the team provides a high-level estimate on each user story and puts the user stories in the product backlog. The product backlog contains the user stories in a prioritized order

external customers. Further sources of requirements can be architects or customer support teams. As a result, the teams will identify specific user stories which deliver customer value. At the beginning of the project those user stories will be prioritized and added to a product backlog. In some cases, the product backlog is owned and maintained directly by the teams. In other projects, product management or the project sponsor is responsible for this backlog.

The team does a rough estimation on the high-level user stories to be able to assess the high-level effort for the project and to get a rough idea of what work can be covered in the short-, mid- or long-term future. The product backlog should be precise enough to provide the team as well as the project sponsor with a good idea of how much can be accomplished in which timeframe. A rule of thumb is to expect that only 50% of the team's capacity will be available to work on these items. The other 50% will most likely be needed to work on unexpected issues or new features that will surface at a later time during the project (Fig. 7.5).

The next level of detailed planning happens at the beginning of each iteration: each team is taking any new requirements that may have come up and updates user stories and priorities. Based on these changes the product backlog will be updated.

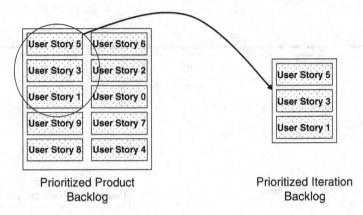

Prioritized Product
Backlog

Prioritized Iteration
Backlog

Fig. 7.6 During the iteration planning, the team takes the items with the highest priority from the product backlog and move them to the iteration backlog, until the team has reached its capacity for this iteration

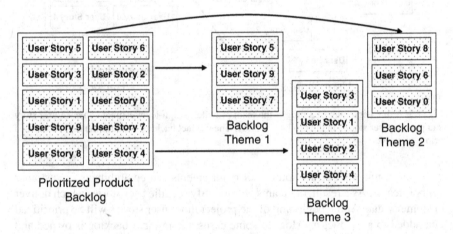

Prioritized Product
Backlog

Backlog
Theme 1

Backlog
Theme 2

Backlog
Theme 3

Fig. 7.7 A larger project that needs several teams to accomplish the work ends up with several backlogs, one per theme

In the next step, a team takes the user stories with the highest priority from the product backlog and moves them into the backlog of the iteration about to start. The team fills up their iteration backlog until they have reached their capacity or velocity for the current iteration.

This is relatively simple if there is only one product backlog and only one team working on the product, but when the product is larger and you have multiple teams working in parallel to get the product done, then it becomes more complicated. Ideally, you would establish teams around features or themes and have them work on these particular themes during the entire project (Figs. 7.6 and 7.7).

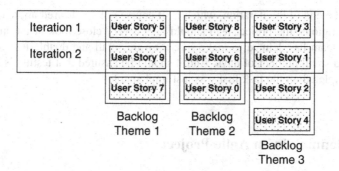

Fig. 7.8 Each team has identified their highest-ranked user story for the iteration

The good thing is that the large problem "product" is now broken into smaller problems per theme, which can now be managed by the individual teams. With only one backlog and one team working on it and that team always taking the user stories with the highest priority from the backlog, you are always sure to provide the highest value in the time available. Now, with multiple teams and multiple backlogs it may not necessarily mean that the complete project team is always working on the items with the highest priority. Let's take the example from the figure above and let's assume that each team can only get one user story done per iteration (assuming that the next user story cannot be done in that same iteration as the total amount of story points is larger than the team's capacity for that iteration) (Fig. 7.8).

That's all very well, but as you can see from the example, the team working on theme 2 is working on user story 8 in iteration 1, before the teams working on themes 1 and 3 are working on user stories 6 and 9, which in the product backlog are ranked higher than user story 8. One possible solution to this would be to have the developers from team 2 (the one that is scheduled to work on theme 2) help with themes 1 and 3 until the first two user stories are implemented and then start to work on theme 3. This may work in this simple example, but during the project you may run into similar situations with every iteration where one team would start working on a user story with a lower priority than those still left in the overall prioritized product backlog. This happens especially if the user stories are relatively small and the teams can accomplish a significant number of them during one iteration. But at the same time you do not want to assemble and disassemble teams with every iteration (or even during iterations). There is a great productivity increase in having the same people working together in the same team during the entire project, or at least for several iterations. There is not an easy solution to this problem, but one needs to at least be aware of this potential issue.

Most teams new to iterations, and especially the time-boxed aspect of it, think that they need iterations that are quite long. If you want to get their buy-in and not start too aggressively, you could start with longer iterations, like 6 or 7 weeks, and shorten them over the period of the project as your team gets used to them. The big risk with longer iterations is that within an iteration a team could start using a waterfall-like approach by first doing the coding and back-loading the testing towards the end of the iteration, which is of course not the desired approach. With longer iterations, the project lead should take a very close look at the individual tasks the team has defined to make sure they are small and doable in a matter of hours or a small number of days. And it should be ensured that testing is immediately completed on the task that was just implemented.

7.6   Planning in an Agile Project

The first step is to get together with the project sponsor and create a joint vision for the project, then write the resulting requirements into user stories that should on the one hand be complete but also as simple as possible. At this time, the acceptance test criteria should also be defined to allow a clean project exit (of course the challenge here is that there is as of yet no exact list of committed content and therefore the acceptance criteria need to reflect this). As the goal is to have a customer-ready deliverable at the end of each iteration, it is paramount to have all the user stories of an iteration completed at the end of that iteration. This is easier to achieve with small user stories that can be completed in a few days than, for example, with only three user stories that would take the whole team and the entire iteration to complete. If one of these three large user stories gets into trouble, then the whole iteration may be in trouble, since stopping a user story during an iteration would mean that the time invested in it was wasted (at least for the current iteration).

At the time the project sponsor and team are defining the user stories based on the product requirement, it may be acceptable to define the user stories a bit less granularly and break one of these user stories down into several at the beginning of the iteration it is planned for.

Each user story should describe a scenario and the functionality that is relevant to the user or the system (Fig. 7.9):

A user story consists of:

1. A short description of functionality and user interaction
2. The business value that it provides
3. The discussions and agreements that happened so far on this user story
4. Tests that should happen
5. Documentation that is required (Fig. 7.10).

Fig. 7.9 Example of a user story

> As an end user, I would like to calculate the value of my stocks based on the real-time value at the Stock Exchange so I can see the accumculated loss or gain.

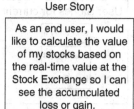

Fig. 7.10 User story with test

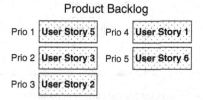

Fig. 7.11 Product backlog

The entire team writes the user stories together. This ensures that everyone has the same understanding and everyone is in agreement. Some teams use small cards to write the story on the front and some keywords on testing on the back. You could also use a simple spreadsheet or one of the software tools on the market.

User stories should be:

1. Small. It should not take more than a few days to implement and test them. Keep in mind that all the user stories selected for an iteration (sometimes also called sprint) need to be completely done (including testing and documentation). And there should be a number of completed user stories per iteration.
2. Independent. A user story needs to stand on its own.
3. Provide business value independent from other user stories.
4. Estimable. The team needs to be able to put a rough estimate on each user story. This also helps to assess which user stories will be completed in the current iteration.
5. And, of course, testable.

Remember – the product backlog consists of all user stories (in a prioritized order) that are not yet implemented:

Now, at the beginning of each iteration, the project sponsor and the team meet for a half-day iteration planning session. During this session the team first looks at any changed or new requirements/user stories and prioritizes them into the product backlog. Then they take the top user stories from the backlog, discuss them to ensure that there is still a common understanding and do a rough sizing (usually in an abstract way using points instead of person days or person weeks). At the end of the session the team has selected the user stories that will be addressed in this iteration (and that fit into the iteration based on the estimates) (Figs. 7.11 and 7.12).

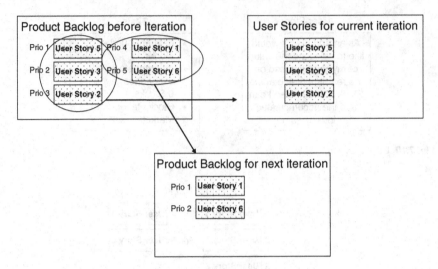

Fig. 7.12 User stories are selected for the current iteration

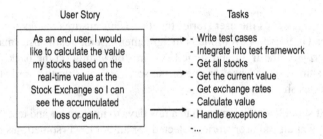

Fig. 7.13 User stories are broken into tasks

Each user story is now broken into tasks. These tasks are sized and assigned to team members.

As part of the selection process, estimates that are as realistic as possible need to be created for each of these tasks. In the next few paragraphs, we discuss the different agile sizing techniques in a bit more detail (Fig. 7.13).

7.7 Estimating

In a traditional (waterfall) project, at the beginning of the project the entire work is broken into tasks using a work breakdown structure. Then all the tasks are estimated. Based on these estimates and the dependencies between the different tasks the work is then assigned to different resources using a Gantt chart (see Figs. 7.14 and 7.15).

The first big difference between a traditional waterfall project and an agile project is that the waterfall project is completely planned upfront. In an agile

	O	Task Name	Duration	Start	Finish
1		Task 1	5 days	Thu 07.05.09	Wed 13.05.09
2		Task 2	3 days	Thu 14.05.09	Mon 18.05.09
3		Task 3	5 days	Thu 07.05.09	Wed 13.05.09
4		Task 4	2 days	Tue 19.05.09	Wed 20.05.09

Fig. 7.14 Gantt chart from a traditional waterfall project

	Task	Previous effort
	Tiles put in regular squares	2 hours per m^2
	Advanced figures	4 hours per m^2
	Tiles put diagonally	3 hours per m^2

Fig. 7.15 Durations for specific tasks on a previous project

project, only very rough estimates are assigned to each user story in the product backlog, and detailed estimates are only done at the beginning of each iteration and only for the user stories that are targeted for that iteration.

In traditional projects, sizings or estimates are usually done in person days or person months. Developers or architects doing the sizings are usually doing this based on their experience and are trying to compare the work ahead with some work in the past to come up with reasonable estimates.

Let's assume you are building a new home and are putting the tiles on the floor and the wall yourself, or helping a friend who is renovating an apartment and wants to have new tiles.

From your last tile project, you remember how long it took you to finish a single room.

You can use this to estimate the durations for your next tiles projects. The big advantage of putting in tiles, especially in a new home, is that millions of people have done this before, and that it is a well known technique. Even building a house or an airplane is something that can be well planned as long as it is not the first time that you are building a completely new type of aircraft. When looking at the aircrafts Airbus is manufacturing, you will notice that most new models are modifications of a previous model. The Airbus A319, for example, is a shorter version of the Airbus A320. You may remember the trouble that Airbus had with its Airbus A380, currently the largest passenger aircraft in the world. This new aircraft was quite different from other aircrafts Airbus had been building so far, one difference being the two decks and its extensive use of carbon fiber materials.

Innovation makes the prediction of schedule and efforts, as well as the anticipation of the challenges ahead of you very difficult.

Late in the year 2000, Airbus decided to build the A380. The design was finalized in early 2001, and production began. But in the end, the first delivery of an Airbus A380 was delayed by over 3 years, as Airbus had several unexpected problems during the manufacturing and tests of the aircraft. The delivery date was moved three times, and the delay caused dissatisfaction with their customers.

Sometimes it is really possible to plan a project quite precisely and to provide reliable estimates: building prefabricated houses, where each home is only a slight variant of a previously built one, is very close to a well defined manufacturing process. Usually, all modifications are known upfront and the number of change requests is very small. Especially since some things are very hard to change after a certain point in the project (like increasing the size of the basement after it is already built). In these projects, it is comparable easy to estimate the cost, the effort needed, and the interfaces between the different work items. You will understand dependencies and the right sequence of the tasks, which need to be completed in the right order (like putting the pipes for the heating in the floor before you put down the tiles).

Manufacturing similar projects may also potentially contain unexpected surprises, but not to the extent likely in a software development project where every project is a new endeavor (otherwise you would just reuse the code). Especially when you are using new and leading-edge techniques, the required efforts of a software development are difficult to anticipate. In software projects, it is usually not possible to have a detailed specification that never changes throughout the project (unless it is only a 1-day project and the sponsor went on vacation after you received the requirements). The estimates you make at the beginning can only be very rough and the team has to learn how much effort it really takes as they go. The risk is smaller the more familiar the development team is with the technologies they are using and the domain the resulting software is used in.

These are just some reasons why agile projects do not even try to size the complete project upfront.

There are two basic practices for estimating agile projects:

- Use an abstract unit and compare the effort for the different tasks, instead of trying to measure in person days or person weeks.
- As you move from iteration to iteration, look back and see how good your estimates were in the previous iteration. Let's say you expected to get 35 units of work done in the previous iteration, but you ended up only getting 30 done. In this case, you should take this into consideration this when deciding what to get done in the current iteration.

7.7.1 Units often used in Agile Projects

As it is difficult to really estimate the effort for software development tasks, there are two main approaches used in agile projects:

- Story points,
- Ideal days.

Story points have no particular unit. They are used to put the different tasks in a relative order with regard to size and difficulty. A task that is more difficult should have more story points than one that is easier.

When using story points it is important that all items are sized by the same group of people to ensure that one story point always represents some sort of same level of difficulty and effort.

In larger projects with several agile teams it is usually impossible to have all teams sizing all items together. Therefore, a more practical solution is to have each team do the estimates on their own, which on the other hand means that a story point of two different teams may not represent the same level of size and difficulty and may therefore not compare. This means that a four-person team delivering ten story points in an iteration is not necessarily better than a four-person team delivering only five story points in an iteration. In fact, the team with the ten story points may have been performing worse than the one with the five story points.

There should be no reason for teams to worry that story points may be used for performance ratings and rankings. They are abstract and the level of difficulty and effort that they represent is defined by each team internally (Fig. 7.16).

Usually the next question is "Why not directly use person days?" or "What is the transformation factor from story points to person days?". What teams should do is record how many story points they were able to deliver in an iteration. Scrum only suggests using the plain number of story points, but we recommend normalizing them with the days invested by the team into this iteration.

In a small project with no customer, a team that only goes on vacation or education together, and where no-one gets ill, plain story points will do the job. But this is not often the case in a real-world project (Fig. 7.17).

The average number of story points delivered by the team in an iteration is also called *velocity*. Taking the velocity of the past iterations helps to assess how many story points the team could deliver per iteration. Based on this past experience, the team can pick a realistic scope of work for the next iteration.

The velocity can also help the team assess how long it will take them to complete the product backlog.

Taking the example from above (Fig. 7.18), velocity can be used independently from the way you estimate the size, time, or difficulty of the user story.

User Stories	Story Points
- Purchase Stocks	8
- Sell Stocks	4
- Authenticate User	2
- Order new password	2
- Calculate account value	3
- Stocks & Investment Fonds details	4
- Signoff	1

Fig. 7.16 User stories with story points

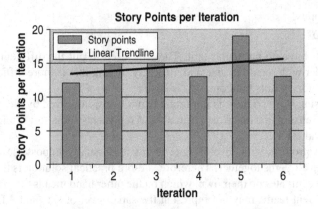

Fig. 7.17 Story points delivered by the team per iteration

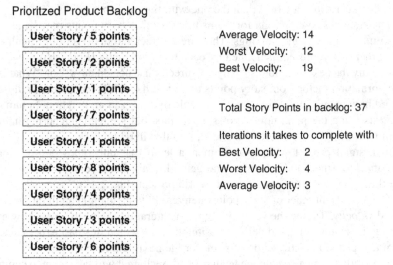

Fig. 7.18 Estimating the time it will take to complete the product backlog based on the average, best, and worst velocity

Ideal person days is another unit often used with agile estimation efforts. Ideal time is:

- The time it takes you to completely finish the user story or tasks.
- The time required to complete design, coding, automated test cases, testing, and documentation and everything else the task requires.
- The time needed for just doing the work, without breaks, meetings, emails, other parallel activities, colleagues asking you for advice, and all the other things that usually stop you from completing the task.

Keep in mind, that if you have an 8-h working day, you usually only make about 4 to 6 hours of progress on your real work, while you spend the remaining time on the other nonrelated tasks. An ideal person day is more than an actual day of a person's work capacity. You'll also find that the "ideal time" to complete a task, once other nonproductive task time is subtracted, is less that you would traditionally expect.

It may make sense for the team to write down all the things they spend time on for the duration of 1 week. Why? Usually, management does not realize how much "nonproductive" time a developer must spend every day in addition to the time spent on actually writing code. Ideal time will also make this more visible than story points. Maybe these are opportunities to become more efficient and eliminate some of these other activities.

If your organization is moving from a traditional development approach with "perfect" plans to agile development, it may be easier to convince people to use ideal days, as they are not too far away from the traditional sizing approaches. Story points are much more abstract and more difficult to assess at the beginning.

7.7.2 Ways to get to Estimates

What is the best way to arrive at good estimates? In the following we discuss some examples, such as:

- Planning Poker,
- Estimate by Analogy,
- Triangulation,
- Disaggregation.

We think *Planning Poker* is a very interesting example.

The idea behind Planning Poker is that the whole team participates in the estimating activity and comes to an agreement on the estimates. Often cards with possible estimates written on them are used. These cards look like poker cards, hence the name Planning Poker. But this can also be done with standard metaplan cards or any other small pieces of paper.

To get the estimates, the team meets – ideally physically in one room. One of the team members assumes the moderator role.

The moderator takes the first user story that needs an estimate and motivates the team members to ask questions. As soon as there are no more questions, the team does a first round of estimates. Every team members decides on their personal estimate and difficulty of user story, then writes down the estimate or chooses the Planning Poker card that matches this estimate (Fig. 7.19).

Then the team discusses these estimates and the team members explain the rationale for their decisions. The goal is to get all the knowledge from the different team members on the table and give everyone insight into the other team members' thoughts. Usually, every team member has an idea on how they would implement

First round of sizings

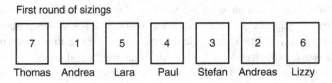

Fig. 7.19 Estimates from a first round of estimates

Second round of sizings

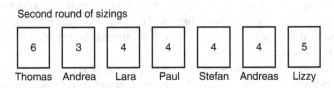

Fig. 7.20 Current estimates after the second round

the user story, and the difficulties envisioned while doing so. The team can learn from each other and arrive at better estimates based on the common knowledge of the team.

After this round of discussion, the team members estimate again (Fig. 7.20):

As you can see, the estimates are beginning to converge. It is now up to the moderator to decide if another round of discussions, followed by new estimates, is required, or to ask Thomas, Andrea, and Lizzy if they can agree to an estimate of 4 for this user story.

One of the advantages is that the entire team agrees to the estimates and buys into them, as opposed an architect or a team lead doing this in behalf of the team. Often the person who did the estimate takes the blame later if it took longer than anticipated.

Planning Poker is ideally done together in one room, but often teams are dispersed around the world, and getting together in one room is just not possible. In these cases, doing a joint planning session and coming to an estimate that is agreed on by all team is still valuable. There is some advantage of everyone turning the cards over at the same time. On the other hand the convergence process can potentially be faster if everyone voices his or her estimates one after the other. In remote teams, you could use collaborative technology that allows all members to enter their estimates and then virtually turns the cards around. You should perhaps make sure that the person who (based on that person's experience or position in the team) is regarded as the thought leader does not "vote" first.

Another approach is what Mike Cohn describes as *Estimate by Analogy*. Here the team looks at previous or other user stories they have already estimated and compares them to each other. A user story that is thought to have a similar difficulty should get the same number of story points, one that is easier should get fewer and naturally one that is more difficult should get more points. We think this is often done implicitly by everyone anyway, as he or she is doing the estimates based on their experience and therefore is implicitly comparing them with other user stories. Planning Poker therefore also contains an aspect of Estimation by Analogy. Estimation by Analogy goes along with *Triangulation*, as you are better off comparing the to-be-estimated user story to two other ones instead of just one.

At the beginning of the release, only a high-level estimate may be required. As you get to the iteration or Sprint planning, you need to become more precise with your estimates. As you break the user stories into tasks, estimate these single tasks and then calculate the estimate for the user story as the sum of the estimates for the single tasks. By doing this, you will get a more precise estimate, as you have gone down another level of detail. This is often called *Disaggregation*.

Whatever the chosen approach will be: It is important to encourage the team members to *not* factor in, how busy they currently are with nonrelated tasks. Estimates are based on story points or ideal person days. They should not reflect how busy a person is at this specific point of time.

7.8 Metrics for an Agile Project

In agile projects it is important to know where the current iteration or the complete project is with regard to the original outlook. A clear goal is to have a version of the product which could be handed over to the customer after each and every iteration. Therefore it is especially important to track the progress of the current iteration to make adjustments early.

User stories need to be 100% complete at the end of the iteration (including test and documentation), stories that are only half done can potentially impact the stability of the complete system and are considered waste as they do not add value to the current iteration. This is another reason why user stories should be complete but also crisp. It should be possible to complete a user story in a matter of days.

Of course a baseline needs to exist before the progress can be measured against it.

One way of tracking progress – as you've seen by now – is using the burndown charts (Fig. 7.21).

The burndown chart uses the same units that were used to estimate the user stories and tasks. Often these are story points or *ideal days,* but any other unit would work too. In the example above, the work for the current iteration was estimate to be 78 story points. Everyday during the Scrum meeting or the daily stand-up meeting, the team discusses the progress each team member has made, any help any of the team members may need, and any surprises that have come up. During this meeting the team also updates the estimates and what it still takes to complete the iteration. In the example above everything went fairly well until May 5th, when the team stumbled over some surprise or at least had to adjust the estimate for one or more user stories. Therefore, the remaining story points went up during these days instead of declining (Fig. 7.22).

Fig. 7.21 Standard burndown chart

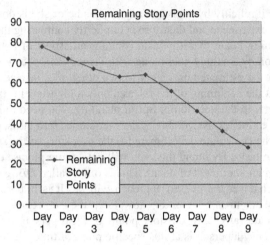

Fig. 7.22 Burndown chart for user stories and tasks

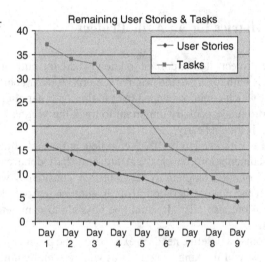

In addition to the regular burndown chart that shows story points or ideal days and does not differentiate between user stories or tasks being completed or not, we suggest to also track the number of user stories or tasks remaining for the current iteration. The purpose here is to ensure that the tasks and user stories are continuously completed including design, development, documentation, and test. Especially in a larger project with several teams where you cannot be directly involved with each team, this additional chart helps to ensure that the team does not revert to a small waterfall model during the iteration, developing and coding all items first and only testing the code towards the end of the iteration. Tracking user stories and tasks is especially useful with longer iterations (like 4–6 weeks) to help ensure that the teams really keep them small and self-contained.

Another helpful chart is the one showing the distribution of the size of the user stories.As you can see in Fig. 7.23, the team has not been able to break down the user stories in parts of about the same size. The smallest user story has three story points, but there are also a significant number of user stories with more than ten story points. Those large user stories will be quite cumbersome to manage. There is no need to have all the user stories have exactly the same work estimate, but it is desirable to have them closer together than shown above.

The team breaking down the user stories in Fig. 7.24 has been able to define much smaller user stories that have similar estimates. This setup has the advantage that the team has enough flexibility to finished the iterations with completed user

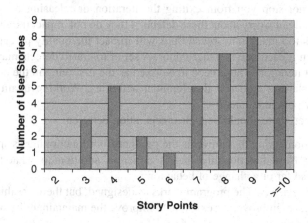

Fig. 7.23 In this example the user stories have significantly varying sizes. There are also many large stories, which will be more difficult to manage

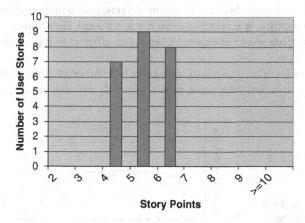

Fig. 7.24 Iteration with user stories that are small and have about the same size

stories (as they are small and there are many of them), compared to a team with only a few large user stories, which always runs the risk that when the iteration is over, one or more user stories are only 90% done and therefore not bringing the expected value or even destabilizing some other parts of the product.

7.9 Defects

There is some discussion going on in the industry on how good defects are as quality indicators. In our opinion, defects are an important indication that we think needs to be tracked and managed. Ideally, the teams should exit the iterations with no defects impacting the quality of the product. Of course, there are always defects that should not stop you from exiting the iteration or releasing the product, as customers will not worry about these defects if they do not impact the functionality. Nevertheless too many unresolved bugs will impact the quality perception of the product, even if each one is rather minor. A set of tolerated defects may be moved into the next iteration. Those need to be fixed as the first thing in the new iteration and before starting any new development activities. Working quality code has priority over new functionality.

In a project, there should be a clear separation between:

- *Defects*: Something that prevents the product from functioning as specified.
- *Features*: New functionality that is either in addition to what the product provides so far or a change of behavior.
- *Refactoring needs*: The program works as designed, but there are things that the team wants to improve, for example to improve the maintainability or to remove unused code.

Separating the refactoring needs from the defects and features especially helps to eliminate work at the end of an iteration or release, because there is then already a clear separation. Teams should exit with no defects, but having refactoring needs at the end of an iteration is acceptable, even if it is not ideal. Refactoring needs are not defects, as the code works as expected, but not as efficient as it potentially possible (Fig. 7.25).

During the iteration there will always be a number of defects in the defect backlog, as it will just take some time to fix a defect and get it into the build.

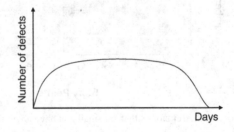

Fig. 7.25 The ideal defect curve that a team should achieve during an iteration

Keeping the backlog of unresolved defects small is important. In all projects, especially in larger ones with separate test teams that do regression, performance, or system testing, it is much more efficient to fix the problems immediately instead of letting the test teams run into the same problems repeatedly.

In larger projects with more than one team or with additional test teams testing in parallel to regular iterations, the team will also get additional defects from these other groups during the iteration. These defects could result from tests under load or stress, or state that performance goals are not yet met. The team has to put some amount of resources aside during the iteration planning to be able to address these defects. If there are more problems in their code than expected, then the team has to delay some user stories in favor of fixing defects.

We have found it useful to manage defects in the development phase (while the teams are still developing code during iterations) differently from the pure release closure/regression testing phase. During the development phase, the defects owned by the agile teams are categorized and managed by these agile teams. Later in the project, during the release closure phase, we found it most effective to manage all defects centrally by a common release management.

It is necessary to monitor the overall defect backlog (categorized in open and working defects, which are owned by the development team, as well as defects to be verified, which are owned by the person or team that opened the defect).

Additional important data is generated by monitoring the defects on an agile team and/or department level (depending on the phase of the project, as explained above). This allows identifying the teams that may need help since their defect backlog is larger than a healthy one would be. If only the overall backlog is monitored, it would be difficult to identify the team that needed help, or the overall backlog could still be in plan, but one team could already be having serious problems while other teams may be doing better than planned, meaning the overall backlog would still look all right. Therefore breaking the defect backlog down to the individual team level is important to identify the teams that need help.

As you can see in Fig. 7.26 below, Team A's defect backlog more or less continuously increased over a period of 3 weeks, finally the team started to focus

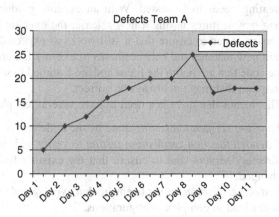

Fig. 7.26 Defect backlog on team level

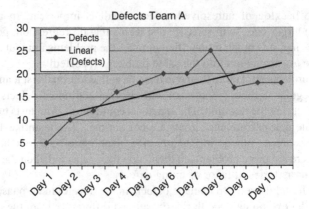

Fig. 7.27 Defect backlog on team level including trend line

more on defects after May 14th and were able to reduce the backlog a little bit, but it is at a high level at the end of the chart. This team needed to put some user stories aside and get the defect backlog under control before implementing new features.

To make the trend more visible, you can also use some additional standard chart tools, like a trend line (Fig. 7.27).

7.10   Independent Release Testing and Wrap-up

Everyone is responsible for testing, and testing is done immediately rather than waiting for a test phase at the end of the project. Test is a continuous activity. Each code change is tested by the development team at the time the code is integrated into the code stream. Test automation is a strict requirement for any development work.

Nevertheless it makes sense to hand over working code to an independent test team on a regular basis for investigative testing of the full product.

If you are starting a new project, then obviously the entire functionality the team is creating needs to be tested. With an existing product, which is enhanced by adding new features during a new release, the existing functionality needs to be regression-tested to ensure that it still works as expected. As a release focuses on certain new features and enhancements to existing features, there are always areas of the code that are not in the focus and are assumed to work unchanged. This is an assumption that may not always be correct.

This means that there is a need to have several test phases:

- *Function verification testing* done by the agile teams during the iterations.
- *Extended function verification testing* to verify additional platforms, databases, directory servers, and to ensure that the existing functionality still works unchanged.
- *System verification testing* to ensure that existing and new functionality performs under load in complex configurations.

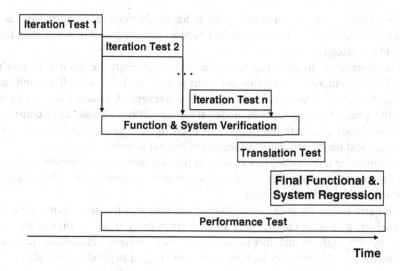

Fig. 7.28 The different release test phases shown on a timeline

- *Translation verification test* to ensure all the translated text shows up correctly in the right place.
- *Performance testing* to ensure that the set performance goals are met.

Figure 7.28 below shows the different test phases on a timeline.

An agile team needs to ensure that the new or changed functionality provided by that team is close to being bug-free. It is important that the teams exit an iteration with ship-ready code, which should mean zero defects (or at least with no defect against which a customer is expected to open a problem ticket) and the first item in the next iteration needs to be the fixing of any remaining defects.

We've already mentioned that the test activities of the agile teams are focused on the code they are writing from scratch or are changing. However, an existing product also has a lot of existing code that is expected to work unchanged. To verify that this is still the case, existing automated test cases for the complete product should be run on every build. We think it is more efficient if this is done centrally for the entire release, instead of having each team doing it for their respective parts. These regression tests need to be carried out on the different environment configurations the product supports, meaning on all the databases, on all the platforms, with all browsers, all directory servers, all the different combinations of these and so on.

This means that a separate extended functional verification test is required that provides all these environments (with the tools to automatically set them up) and executes the regression tests continuously throughout the project duration. We found it more efficient to have the agile teams "just" cover one platform and database type and have variations of the system environment tested by this central extended function verification test team.

There is also a specific phase during the extended functional verification test that we call globalization verification test. Even if the product were not translated and delivered only in English, it must be able to handle data entered in other languages, which may have other characters (like Chinese) or may be written and read from right to left (like Hebrew).

The system verification test takes the drivers at every iteration exit, sets up complex, customer-like scenarios, and runs tests to do load- as well as limit- and boundary-tests with large volumes of data over several days and weeks to ensure that the product is stable even under these conditions. These tests consist of regression tests to ensure the stability of unchanged features, and of new tests to explicitly test the functionality added in the current release.

Performance test should start as soon as the first iteration is complete, to ensure that the existing performance requirements are met and that the next features meet their respective performance goals.

All these necessary tests, in addition to the tests that the agile teams are doing themselves, also help uncover any shortcomings (may there be any) of the tests done by the agile teams themselves. Especially system verification tests and performance tests have very high demands with regard to good code quality.

7.11   Involving Customers

In the chapter on planning, we stated that agile project management starts with defining resources and dates, but only does rough requirement estimates. Even the exact list of delivered content will be kept open until the very end of the project (see Fig. 6.1).

The bad news is that a customer will probably never accept a software deal in which he has to pay for a deliverable without having a commitment about its content.

Is this a fundamental flaw with our agile approach?

No!

First of all, the waterfall approach doesn't help much either: While the requirements are fixed, the delivery date and the required resources are just estimates (and guesses?), which isn't much better.

Let us face this: In software development you will never have certainty on all three variables of that magic triangle upfront. There will always be risk. This risk can be taken either by a customer, if the price is set based on actual effort, or by the software developer, if the price is fixed.

In either case, yet again communication, collaboration, and trust will be the essential foundation of success for both customer and developer. Indeed, both are equally vested in success, as only a relationship that is profitable for both sides will be sustainable on the long run.

Involving customers early and directly can make a tremendous difference to the project. Keep in mind that good quality does not only imply absence of bugs, it primarily means fulfillment of the customer's expectations.

Even the customer may change their mind about the desired content. It will also be of advantage for the customer if the scope of work is not set in stone at the

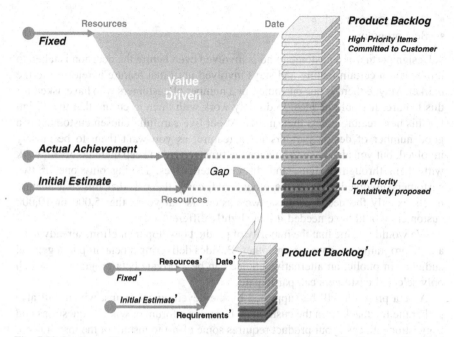

Fig. 7.29 Dealing with requirements in customer engagements: Once the time-boxed project duration ends, a follow-on project, or an additional iteration can be appended to cover further requirements

beginning of the project. He will definitely have an opinion on prioritization and may not even care if less important items are dropped.

If the significant requirements cannot be achieved at the end of the project duration, there is still the option to append another iteration or even kick-off a follow-on project. In this context it will become extremely relevant who agreed to cover the risk of any gaps between the actual progress and the initial estimates that were part of the contract (see Fig. 7.29).

Software development projects that are directly developing a solution for one customer should ensure that the customer is directly involved from the very beginning and stays involved over time, which is often the case anyway. Every iteration end is a good time to provide the customer with a working version of this new solution. It would be ideal if the customer had an architect with domain knowledge who directly worked with the team on a daily basis and sat on the same floor as the development team to ensure instant collaboration in case of detailed questions about a certain requirement, or in case design decisions have to be made. In cases where this is not possible, it would be good if the customer were involved in the daily scrum meetings (in case there is just one team working on the project) or had a daily interlock with the Scrum Master to discuss any issues that may have come up.

If you are in the business of developing a product that is or will be commercially sold on the market, then we would recommend the following two basic concepts:

- Design partners and
- Beta program.

A Design partner is a customer who is involved even before the decision is taken to implement a certain feature and stays involved until that feature is released to the market. Maybe there is one customer, or a number of customers who have asked for this feature. It would be ideal to directly work with them to ensure that the design for this new feature meets their needs. About five carefully chosen customers is a good number of design partners for a feature, as you want them to be closely involved, but you also have to balance the time needed to keep them involved, walk with them through designs, and discuss alternatives. Having only one or two customers involved may on the other hand, carry the risk that this new feature meets exactly the needs of these two customers but your other 5,000 or 10,000 customers would have needed it in a slightly different way.

We would assume that the majority of product development efforts already have a Beta program in one way or the other. Besides delivering a beta drop to a general audience in public, an alternative choice can be a private beta program, in which only selected customers can participate.

A beta program will be supported by the development teams, which will also gather the feedback from the customers. Useful is a forum or wiki for questions and suggestions. In case your product requires some effort to install, or the install is not completely finished yet, then you could also just provide images or the installed products using one of the virtualization products on the market, like VMware®.

Chapter 8
Mix and Match

8.1 The Tragedy of Being Successful

Typically, in startup companies without a strong hierarchy, a small group of people kicks off projects quite informally. They simply begin to work and continue to evolve their ideas towards a common vision. Successful startups gain competitive advantages because they are able to adapt to customer needs in a nonbureaucratic way. They follow a lightweight, ad-hoc development approach and avoid too large a burden of processes, documentation, management control, or project planning. They operate extremely cost-effectively and focus on implementing customer value as quickly as possible, in order to generate some return of investment. The progress can be impressive, as the most relevant and attractive use cases will be implemented first.

Teams are rather small, everybody knows each other, and everyone is in it together. People in startup companies also tend to have a strong passion for their work and show an entrepreneurial attitude. The future of each team member is strongly tied to the success of the project.

Over time, the product they are developing may be a tremendous success.

Things will change in such a startup company once the team starts growing rapidly. Different teams will be established to focus on different topics. This work-split happens naturally, without a master plan. Maybe there will be goals that define the overall direction. And maybe everything we wrote about management by objectives and self-organizing, horizontal teams will be nicely implemented in that company.

Over time, the customer base may be prospering.

But by now, the easy and most rewarding use cases are already implemented. Customers are now asking for more complex things, like high-availability or much better performance. Once the low-hanging fruits are gone, there will come more specific use cases, which may possibly be added for only a few influential customers.

Over time, the code base will become large and complex.

Maintenance costs will increase. Troubleshooting will be difficult, as not every programmer who once wrote pieces of the code is still around. Architectural limitations of the initial design are being encountered now, and rework becomes necessary in several areas of the code. It is obvious that the ad-hoc development mentality applied in the early days of the startup company cannot go on forever. Customers expect reliability and well-defined patterns for interaction (e.g., as part of a requirement process or bug reporting).

Over time, management, rules, and directives will be introduced.

A split of work is beginning to establish itself to solve specific kinds of tasks in an optimized manner. To service the customer base, a support process will be established. Guidelines will be issued to set internal coding standards and best practices. Additional platforms may need to be supported to convey the perception of completeness, to comply with internal policies, or to make an important deal. Customers may mandate backward compatibility of published APIs, or ask for extra work to support accessibility standards and translation requirements. Comprehensive checklists, reviews, and approvals for designs will be introduced. Securing intellectual property and other legal issues will become a focus area. Sales teams, solution engineers, consultants will be introduced to serve as a link between development and the customers. Often such measures are added for the best, e.g., as the result of lessons learned and to avoid mistakes in the future, or due to the need to manage a much larger customer base efficiently. Nevertheless, while individual aspects may be perfected to some extent, the overall development process will become more heavy-weight, communication needs and logistics will increase. In addition, more and more developers are tied up in customer support and bug fixing activities and in extra work to ensure the product's compliance with imposed rules and standards.

The progress and pace of innovation will become sluggish.

Inevitably, the competitive advantage of the early days will be lost to some extent. The pace of innovation will slow down.

These observations do not only apply to startup companies. Projects within larger enterprises can show just the same patterns as they evolve from small "stunkwork" activities to successful products that require a mature development organization to support a larger number of customers.

8.2   About WebSphere Portal

One real-life example of a development organization that has reached maturity is IBM's Internet and Enterprise Portal product, WebSphere Portal. It began as a small project, driven by a dozen colocated developers. Everyone could still oversee pretty much any area of the code base. The lead architect and team lead were among the active programmers. The first release was more or less a joint project with a single customer. To get started quickly, the implementation had been based on an open source component. During the course of several releases, the product became more powerful and extremely rich in terms of functionality, reliability, and scalability. Today, analysts classify WebSphere Portal as the market leader in its business space, and its customer base contributes significantly the IBM's overall revenue. The product has become the strategic web user interface in IBM's Service Oriented Architecture.

Meanwhile, WebSphere Portal is being developed by several hundred employees in various locations around the globe, and the organization services a large customer base.

The Portal project was managed following a classical waterfall approach. The development plan of a single release spanned a duration of more than a year. Product owners and project management had to deal with a huge amount of long-term plan content, as more and more additional requirements came back from the growing number of customers. At the same time, more and more developers were kept busy on short notice with customer support activities and maintenance work. Juggling all these work items on the one hand and on the other having a fixed release plan that covered over a year became more and more of a challenge. The planning horizon was too far off, and the level of detail which the centralize release management had to deal with was far too fine-grained.

When planning that far ahead, the creation of a project schedule is difficult and time-consuming. The sizing estimates used for planning were unreliable because they covered too long a period of time, with a lot of ambiguity in between. How much buffer needs to be allocated? How do you get all stakeholders to agree on the proposed plan content and how can project management commit to the ability to deliver precisely what has been negotiated, based on extremely rough long-term predictions and without thorough understanding of the details? Most often, developers had already started implementations on their own account, while the overall release schedule was still being debated between project management and product owners.

One astonishing observation has been that several key features did grow outside of the actual development team. Intern students, which are not part of the project planning activities, turned out to be an extremely creative source of innovation.

When, over time, planning became a nightmare and the organization began to operate too sluggishly, the time for simplification had arrived:

The Portal organization decided to move towards an agile development approach. The goals were to return to the vitality the project had shown in the beginning, to increase competitiveness, and to defend the market leadership against other newcomers.

The questions to be answered were: Is such a turnaround possible at all? How can a project culture be changed from a plan-oriented model towards a more agile approach? Is such a large project able to follow agile development practices at all? Does agile thinking scale beyond small research activities? Let's recapitulate the major ideas behind agile development and see where, how, and to what extent they can be and were applied to a large project such as WebSphere Portal.

8.3 Which Projects are suitable for Agile Software Development?

Probably the most common excuse for not moving towards an agile development model is a statement like "we are different, agile will not work for us." But which projects are suitable candidates for implementing an agile approach? The variety of answers to this question is likely to be overwhelming. When attempting to sort them out, there are three aspects to consider:

• The characteristics of the organization in which a project is executed,
• The nature of the project, and
• Its size.

The first of these aspects is probably the most critical one: Companies that emphasize regulations and a strict order of "how things are done" will always have difficulties adopting agile software development, regardless of the particular project you are looking at. Agility will only work if the organization is "lean" and its leaders are ready to take the risk of letting go of control. But not only the leadership team needs to support the move to agility: you cannot force an agile process on the development teams either. Teams need to be picking up the responsibility and need to start thinking as stakeholders. An organization that already has experience with agile software development will be able to benefit more from an agile approach than an organization that is just getting acquainted with agility for the very first time.

The nature of the project is another important aspect: Projects with ambiguous or even unclear constraints will definitely benefit from adding agility. A first-of-a-kind project that explores new technologies, the development of an entirely new product, or any sort of research activity are a perfect and easy match for agile techniques. Larger projects or the evolution of existing solutions are also good candidates for agile, but require some more customization of the general agile processes.

Qualification of the team is another factor: A staff of experienced and senior developers will make the move towards agile much easier. A majority of lower-skilled or junior team members will likely increase the need for fine-grained planning and guidance in many details.

Extremely mission-critical projects with a high demand for documentation, auditing, and control will rarely be good candidates for running the project based on agile methodology. The same applies to projects that need to meet confined regulatory requirements to comply with standards and certifications, such as the Sarbanes-Oxley Act.

Distributed teams and off-shoring of work will make any project more difficult to manage. We do not believe that agile practices will add to these difficulties, nor that they will make virtual teaming significantly easier, although management by objectives, self-organization, and decentralization are definitely good approaches for working with remote teams. This is especially true if you can break the project into smaller themes or features and assemble teams that are colocated to work on these separate themes.

And finally, project size is relevant when considering agile software development. While it is obvious that small projects with a dozen team members are extremely well-suited for agile development processes, opinions on the maximum team size agile methods can cope with vary greatly.

There is in fact much doubt that agility is able to manage larger projects. Nevertheless, we are convinced of exactly the *opposite:* in particular large projects with considerable complexity deal with tremendous planning challenges and tend to waste time and resources on bureaucracy and micromanagement. A disciplined agile software development will prove extremely helpful in simplifying project management and dramatically improving the time to market.

We do agree that setting up an agile environment for large projects can turn out to be something of a challenge. The techniques and concepts described in the previous chapters will help you do so. Be prepared and keep in mind that it will take a strenuous transition period until a large project will actually benefit from agile software development.

8.4 Scaling Agile

Managing large projects with agile methodology can significantly simplify project execution. To be provocative: By letting go of control and giving up centralized project management, the complexity of a large project can be better controlled.

All techniques and practices we have discussed so far actually scale very well: short iterations and small pieces of work, management by objectives, teams that assume end-to-end responsibility, a powerful collaboration infrastructure, and so forth...

Let's recapitulate the main concepts that help scale an agile project:

- *"Scrum of scrum":* The concept of Scrum postulates daily meetings to exchange the most recent status of a team. A scrum meeting should have a small number of participants to keep the meeting short and useful. To make this scale for a larger project, *scrum of scrum* is introduced: an additional meeting that is attended by one representative of each scrum meeting. Each of them delivers a consolidated summary of their own scrum meeting. They provide information, exchange feedback across teams, and jointly define the status of the overall project. Scrum of scrum is basically the idea of establishing a hierarchy of project status meetings.
- *Continuous Integration:* In large projects, where many developers produce a significant amount of code, it is crucial to integrate results as quickly as possible into a common code stream. Delaying this integration can easily cause functional dependencies to go out of sync. Concurrent code changes can interfere with each other and cause build breaks. Tedious integration work may then become

necessary to align code that has been developed in parallel on separate streams over an extended period of time. Integrating new code continuously ensures that integration issues are discovered early and can be resolved before they begin to accumulate and cause a general instability of the entire system.

- *Build Infrastructure:* The larger the code base, the more important the efficiency of the build process that turns source code into installable products. The build infrastructure needs to pull in specific versions of components and prerequisites from different locations. Builds should be executed frequently (daily at the very least) and should not take long.

- *Test Automation:* Once code is integrated continuously and builds produce new versions of the product frequently, regression tests must be run on a continuous basis. This is only possible if they have been automated to a large extent. A suite of automated test scenarios is instrumental in giving quick quality feedback on each new completed build. In addition, it is important to increase test coverage to ensure that new code did not break existing functionality.

8.5   Moving Towards Agile

Introducing agile software development is without any doubt a challenge – especially for teams operating in a plan-driven waterfall approach. An agile approach is not only a new method to apply. It also requires fundamental rethinking of mentality, culture, processes, and behavior. It affects software architecture, team organization, leadership style, and project management. Such a radical shift is a scary thought for many companies.

To get there, many things need to change.

Introducing such dramatic changes into an organization is not a trivial thing.

It is unlikely that this will happen as part of a "revolution," where the teams and management decide to change the development approach overnight. It will rather be a journey of small steps over time, with agile patterns such as autonomy of teams or planning less far ahead slowly evolving. The organization will need to make the necessary changes step by step, project by project, team by team, and will eventually reap the benefits of agile software development.

It is advisable to get started quickly and learn as you go, rather than waiting for every question to be answered and every issue to be addressed. Start small and focus on the low-hanging fruit, solve the most pressing issues first, and find the fit best suited for your project.

A pragmatic view on transitioning to agile software development can be described as *"crawl – walk – run."*

- *Crawl:* At the beginning, changes in project management are rather insignificant. Most of the plan-driven legacy methods are still in use. Teams still stick to an overall schedule and measure progress against targets defined at project kickoff. But gradually, agile elements are being introduced, such as defining a backlog list for each team, working in iterations, and continuously integrating code. The change of mindset takes time, however: during the iterations, developers still focus mostly on writing code, and testing still happens predominantly at the very end of the project. While the buzz phrase of "agile software development" is in the air, the distinct phases of design, code, unit test, and test can still be clearly observed on the ground.

- *Walk:* Flexibility increases with a shrinking planning horizon, processes are slowly being simplified. Teams begin to translate their goals into use cases without being prompted. They begin to manage their iteration backlog on their own account. Stability of the code increases as developers put more focus on extensive testing. Working and tested code is now clearly accepted as a priority over working on new features until the last minute. Agile thinking begins to pay off, as fairly stable code is available for hand-over to customers significantly earlier than before. Nevertheless, not everyone is convinced yet. There is a strong desire to clearly specify in advance the exact content of the product to be developed. Changing the schedule, plan, or content of a product is considered disruptive and unsettles the teams. There is still an extensive test phase after the last iteration, since agile is not yet fully trusted and there is a strong desire to play it safe.
- *Run:* Experience and self-confidence grow. Over time, project management dares to let go of control to a larger extent. The overall project or release goals with the high-level user stories are defined at the beginning, but the detailed planning and execution focuses on the current iteration. Teams accept that the future beyond the current iteration will be discussed and decided when the time comes. The continuous rhythm of iterations and proceeding step by step shapes the mindset of the organization. The project becomes largely agnostic to overall changes in direction. The anxiety about the less guided and less deterministic development approach settles.

A good method for assessing how mature your agile development approach has become is to monitor how the amount of unresolved defects changes during project execution. If there is a strong emphasis on resolving the accumulated list of bugs towards the end of the project, the adoption of continuous integration is not yet successful.

In a mature agile development project, the number of unresolved defects will be a rather flat curve without any significant peaks. Meaningful testing can be started much earlier, as there are fewer unresolved critical issues that block the progress of the test teams. Furthermore, performance analysis and complex load testing can be started early, so that there is enough time to react to the identified issues.

Transitioning towards agile means shifting test activities more towards the early stages of the project, while development activities are continuing until very late in the game.

Fig. 8.1 shows the characteristic curves of unresolved defects and how the development and test efforts are distributed along the timeline of a project.

But regardless of the speed of transition and in which phase you currently are, the move to agile software development will only be successful if upper management

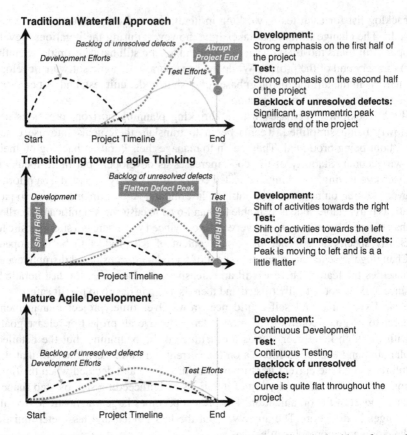

Traditional Waterfall Approach

Backlog of unresolved defects

Development Efforts

Abrupt Project End

Test Efforts

Start   Project Timeline   End

Development:
Strong emphasis on the first half of the project
Test:
Strong emphasis on the second half of the project
Backlock of unresolved defects:
Significant, asymmentric peak towards end of the project

Transitioning toward agile Thinking

Backlog of unresolved defects

Flatten Defect Peak

Development   Test

Shift Right   Shift Right

Project Timeline

Development:
Shift of activities towards the right
Test:
Shift of activities towards the left
Backlock of unresolved defects:
Peak is moving to left and is a a little flatter

Mature Agile Development

Backlog of unresolved defects
Development Efforts

Test Efforts

Start   Project Timeline   End

Development:
Continuous Development
Test:
Continuous Testing
Backlock of unresolved defects:
Curve is quite flat throughout the project

Fig. 8.1 The distributions of development and test efforts, as well as the backlog of unresolved defects, are good indicators of an agile development approach's maturity

supports it wholeheartedly, and if the development teams pick up the responsibility and drive their assigned goals forward on their own account. Clear, obvious, and continuing executive support as well as training, mentoring, and coaching will support the transformation of the project management style and predominant team mentality. Involve people and encourage them to contribute with their experience and practices. Keep in mind that a lack of acceptance within the team, resistance towards change, or interfering too much with the team activities are showstoppers for becoming more agile.

During the transition phase, it is important to keep the overall vision in mind. What are the most urgent pain points to resolve? Which particular aspects of the development process should agile practices optimize? Keep in mind: agile software development is not just a method or a set of rules to follow. Agile thinking comprises many different methods, techniques, ideas, and tools. Each project needs to choose from this set of options and should use those that best fit their needs and constraints. A specific agile development process will always be an individual mix and match of known methods, and it will grow and evolve. Specific practices can be learned, applied, and evaluated individually and independently, at any given point in time and at the pace that suits best.

Part of the transition towards an agile development process will be reflections: Hold regular and frequent lessons-learned meetings, but keep them short and time-boxed! The collected feedback and experience helps to find the right balance between flexibility and structure, autonomy and guidance, commitment and ambiguity, documentation and hands-on prototyping: Reflection will change your individual implementation of agility over time! Remember: the journey towards a truly agile development process will be a continuous endeavor.

"No man ever steps in the same river twice, for it is not the same river." (Heraklit)

8.6 Tiger Teams in WebSphere Portal

We have seen that agile software development is indeed applicable to large projects, and that the transition from a waterfall approach to agile software development is possible and desirable. Individual projects will mix and match different methods to find a customized agile process best suited for the particular needs of the teams.

Let's go back to WebSphere Portal and have a closer look at this development project's implementation of an agile development process. In Portal's own interpretation of agility, the term *"tiger teams"* was introduced for the cross-functional teams that drive use cases end-to-end following agile considerations. The task of these tiger teams is to foster intensive collaboration across organizational structures, geographical locations, and functional components. Tiger Teams also combine developers and testers. And most of all they work at lowering the center of gravity to enable better, quicker, and more optimized decisions. The goal is to close on final release content late in the planning cycle, while at the same time starting to make valuable progress in focus areas as early as possible.

The Portal team emphasized the following four key concepts in particular:

- *Budget- based prioritization and content definition:* Requirements and use cases
- *Cross-organization teaming structure:* Tiger teams drive innovation
- *Iterative development approach:* Evolve a solution and avoid the "Big Bang"
- *Integrating test and development:* Gain trust and quality from day 1

The following section detail how these key concepts have been implemented in WebSphere Portal.

8.6.1 *Budget-based Prioritization*

Which features should be included in a release? Who decides the release content with which level of granularity? For Portal, the executives, product management and lead architects come up with a first cut of release content by aggregating customer feedback and high-level requirements into rough focus areas. Each of these focus areas is associated with a budget that reflects the approximate number of developers assumed to be working on that area throughout the upcoming release project. In Portal, those developers and testers who will pursue the same focus area are grouped into a *"tiger team."* Management acknowledges that the members of each tiger team are the subject matter experts, and that the "center of gravity" is lowered to allow for quicker, better, and more optimized decisions directly by the team.

While a significant amount of responsibility for the exact content is delegated, the budget remains the key instrument used by the leadership team to influence the focus of a release project: The initial budget defined for a tiger team is only a first rough estimate to help the tiger team with some guidelines. Throughout the project this budget will be adjusted to ensure that the most important focus items are done with a solid staffing and all necessary skills. Increasing or reducing the size of a team increases or reduces the amount of work, which can be delivered for a particular focus area.

This distributed approach scales much better than that of a central release management that decides on all detailed use cases of the entire release and maintains a complex overall project plan for all worldwide developers.

The developers in each tiger team are responsible for translating the high-level requirements into specific use cases. Each tiger team prioritizes the use cases it intends to deliver in a *team charter* document. The team charter outlines the mission of the team and most of all, it lists its product backlog. The product backlog focuses on implementing the team's focus area to the complete scope, but since not everything may fit into one release, use cases are prioritized.

While the tiger teams formally own the team charter documents, it is absolutely essential to engage the leadership team, customers and marketing representatives, when defining and prioritizing the product backlog (Fig. 8.2).

Usually, the market demands certain features at a certain time and even an agile plan has to commit to a minimum of enhancements early on in the cycle. To leave sufficient room for agility, it is extremely important that each

The Team Charter:
- The team charter is the *mission document*, which is used to justify the existence of this team. It outlines requirements, vision, strategy, high level goals, and usage scenarios.
- It is the contract between the tiger team and the *management, customers and marketing*.
- The charter includes a *prioritized list of all relevant use cases* ("product backlog"). Each use case should be implementable and testable within one iteration.
- The charter includes accurate and up-to-date names of all relevant stakeholders and other people who are impacted (i.e. dependencies). This distribution list is used as a baseline when socializing designs.

Fig. 8.2 The team charter

team commits only use cases up to a limited amount of its team capacity. The charter lists use cases that have high priority, and which the team commits to deliver under all circumstances. Other items have the disclaimer of being tentative, and will be implemented in the order of priority as time permits. Some items are listed for completeness, but are marked as out-of-scope for the current release project. Often, the final scope and timeline of the release project are not yet fully defined at the time the team is established.

Each team continuously updates and adapts their charter document as well as any other planning document to reflect changing overall constraints, like modified budget, schedule, and priorities. Dependencies on other teams or different high-level requirements will also affect the charter.

A tiger team is inaugurated to drive a cross cutting theme. Within the constraints of staffing budget, overall release timeline, and given high-level requirements, the team can begin planning and executing activities autonomously. Often, a tiger team starts its work independently of a specific release and drives towards early prototyping outside of an official product. At one point the team's activities will fold into a release project. From that point onwards, the deliverables will begin to continuously integrate into the code stream of that designated release. The lifetime of a tiger team can span one or more releases. The team will be abandoned as soon as its mission is accomplished.

For example, the "Web 2.0" tiger team had the mission to enhance the user experience of WebSphere Portal by applying Web 2.0 concepts. The team was inaugurated long before any planning activities for a particular release took place. Without knowing yet when and how their code would eventually be delivered to the market, they shipped early prototypes and presented them at the LotusSphere conference, on IBM's greenhouse site, and as part of the Portal 6.1 Beta. But most importantly, the customer feedback they received was continuously used to further improve the Web 2.0 capability throughout the project.

After 5 months of development, a change of the underlying web application server has been decided to meet customer requests. This modification affected design documents and implementation significantly, since this new version introduced an entirely new user management component and had different non-functional requirements. The project duration had to be extended. Therefore the team added more use cases to their charter document and spent the extra time until shipment on further development activities.

Finally the achievements of the Web 2.0 team were delivered to the market as a highlight of WebSphere Portal 6.1. Fig. 8.3 illustrates how this tiger team navigated within its scope under changing constraints.

8.6.2 Cross-Organizational Teaming Structure

The Portal overall organization has traditionally been structured into functional components. The management and reporting hierarchy reflects this by having

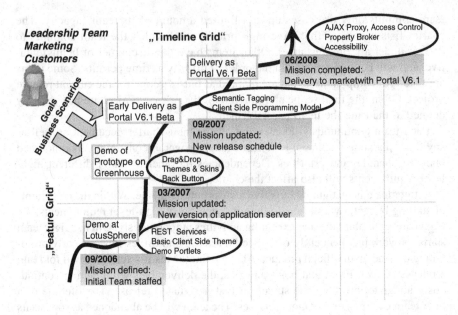

Fig 8.3 Navigating within changing constraints

dedicated teams focusing on particular components, such as administration user interface, aggregation engine, content management, security layer, and so on. These teams work in seven locations around the world: the administration user interface team is in the United States, for example, the aggregation engine is being developed in Germany, and the content management team is located in Australia.

Typically, the development of a new feature involves multiple components. In the past, we had a strict component specific code ownership. A smart project plan has been elaborated for the entire release, which attempted to reflect all the functional and timing dependencies between the different teams. As you can imagine, this has always been an extremely difficult and error-prone thing to do.

Portal's adaptation of agile software development did not change the way the reporting hierarchy is structured into components. Based on his specific expertise and his knowledge of the code base, each individual remains to have a home within a particular component team. But we have extended the existing hierarchy to a matrix organization: the new tiger teams were formed to add a holistic view of overall use cases. They are focusing on their end-to-end deliverable rather than on existing organizational structures. They are tasked to think across component boundaries and to keep the integration aspects in mind. Each tiger team is planning, designing, coding, documenting, and also testing their use cases by themselves, rather than just coordinating multiple parallel code changes done by other teams. This approach minimizes the cross-team dependencies, hand-offs, and avoidable task switching. With the introduction of tiger teams, we changed the mindset of new development activities to a much broader view.

For each new focus area of the release such a tiger team is recruited from the various component teams in order to bring the right set of skills together. These selected experts will collaborate as a virtual team until their mission is accomplished.

Most importantly, the functional verification testing of the team's deliverables is the team's own responsibility. Therefore testers are part of tiger teams as well. Continuous testing and direct collaboration between developers and testers improve the process of troubleshooting and bug fixing tremendously and create a very efficient bridge between the development and test organization.

Ideally, tiger teams consist of members which contribute all the necessary skills to implement the new features. Most of all they would have knowledge and experience to make all required changes to the involved components by themselves. In theory, everyone can *equally* touch any code to fulfill his mission as we have moved towards a collective code ownership. Obviously, this is not always possible. In reality there will always some component specialists which will be *"more equal"* and have deeper insights into a specific area of the code.

This is why tiger teams are just one dimension of a matrix organization. On the other dimension, the component teams are still required. Individuals are assigned to tiger teams in order to drive new cross cutting themes. But at the same time, they keep their strong affinity to a particular component, based on their very unique skills. They will be the experts who know their components in great detail and can contribute their knowledge in design discussions and code reviews. Whenever others touch their code, they are the decisive gate keepers to govern consistency and the correct usage of their component's APIs. In addition, component teams own the maintenance of their code. For instance, they assist the support teams with customer problems and they will address bugs or regression problems, which are outside of the scope of tiger teams.

While tiger teams are rather transient, component teams are long-lived. They contribute significantly to the consistency of the code base and are the places were skill development occurs. In our first release using agile tiger teams, we under-estimated the work still left to the component teams.

The two dimensional matrix of component and tiger teams is depicted in Fig. 8.4.

Having colocated teams would be ideal, but in our project the components are spread all over the world and therefore the required knowledge is not available in one location in all cases. Thus it is unfortunately a given to have teams with members working in different locations. If the travel budget allows it, then we would recommend having the teams meet in person for a few weeks at the beginning of the release, to get to know each other better, do the initial estimates together, and agree on the basic approach.

8.6.3 Evolving the Product in Iterations

It is difficult to integrate a significant amount of code at a certain milestone date without causing a major disruption. To avoid a painful integration struggle, Portal

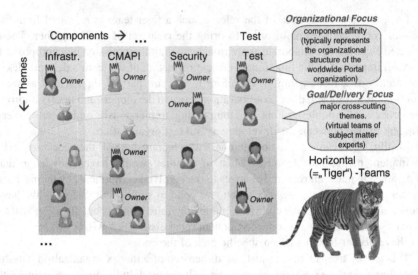

Fig. 8.4 Teaming: theme and component matrix

has adopted an iterative development model. Code is continuously integrated into a common code stream. Functionality is brought forward in several small iterations guided by goals and priorities set by the leadership team. Fig. 8.5 illustrates this approach.

The purpose of an iterative development is well summarized in IBM's Agile Manifesto:

"Agile software development uses continuous stakeholder feedback to deliver high-quality and consumable code through use cases and a series of short, time-boxed iterations."

This implies a few key assumptions:

- Throughout the release, the teams maintain their *team charter* document, which includes the prioritized product backlog with all use cases they tentatively want to address in the foreseeable future.
- They elaborate a rough *high-level design* outlining all items of their focus area (Fig. 8.6). Only the current iteration is being precisely planned, confirmed, and detailed into an iteration backlog, which lists the low-level use case descriptions.
- The *duration* of an iteration varies between 4 and 6 weeks.
- Iterations are *time-boxed*. They have a defined start and end date. Usually, all tiger teams operate on the same iteration schedule.
- The *content* of each iteration is defined at the beginning of each iteration. A tiger team picks the top use cases from the prioritized product backlog in their team charter and starts designing and coding those items.
- Large user stories are broken in to into *smaller, digestible chunks* to ensure that a use case can be implemented within an iteration.

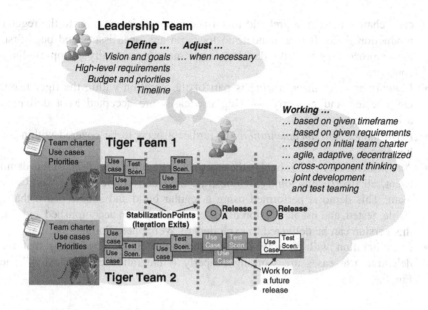

Fig. 8.5 Running a project in iterations

High Level Design Document:
- Provide a technical description of the usage scenarios.
- Emphasize key architectural decisions.
- Spell out implementation dependencies between teams.
- Provide a baseline for further design and implementation work.
- Ensure architectural consistency.
- Deliver a starting point from which further documentation can be evolved.
- Outline required test cases.
- Include presentation material and collateral which helps to explain the deliverables

Fig. 8.6 High-level design document: keep it short and meaningful and share with the right audience in time

- The teams *continuously integrate* their code throughout the iteration, documentation, and automated test cases into a common code stream. There are daily builds of the entire product. Continuous integration with immediate testing is done to avoid destabilization.
- Ensuring the *stability* of each build is everyone's responsibility. Disruptive changes must be avoided by all means. Every single developer will plan and perform thorough unit testing and automated regression for the code he is delivering. The build environment gives some support by running each

code change through a prebuild to surface potential issues prior to the regular production build. It is a mandate to focus on any open issues and bugs first, before proceeding with the development of new functionality ("Stop-the-line" concept).

- *Functional verification testing* is part of the iteration within the tiger team. Only tested and properly working use cases are accepted as a delivered achievement.
- *Performance* and *documentation* are further aspects to be covered within the iteration.
- An *"Integration Fest"* is held at the end of each iteration (Fig. 8.7). At this demo event, all tiger teams jointly demonstrate their deliverables to the worldwide team. This demo is performed using a regular build and should prove that a stable, tested, and usable version of the product has been accomplished and that this version can be delivered to exploiters of WebSphere Portal.
- Each iteration will be *signed-off* by the stakeholders, confirming that the delivered use cases are working properly ("integration exit checklist," see Fig. 8.8).

Integration Fest (Show and Tell):
- Present working code at the end of each iteration to all stakeholders
- Use common demo machine and driver for all teams
- Use the official build without patches
- Do only show a live demo - no PowerPoint charts
- The integration fest happens before iteration exit and is part of the team responsibilities
- Demonstrate working results and share information about the progress
- Explain technologies and techniques, why they are used and what they implications are
- Provide a stable baseline for the next iteration
- Ask for quick feedback from internal or external customers

Fig. 8.7 Integration fest

Iteration Exit Checklist (Quality Certification):
- Is the quality of delivered use cases acceptable?
- Is there a sufficient coverage of automated test cases?
- Has the functional verification testing completed?
- Have the use cases been demonstrated at the Integration *Fest?*
- Is end user documentation available?

Fig. 8.8 Iteration exit checklist

- Independent *validation and verification* of the delivered use cases follows the completion of each iteration. This involves performance and system testers as well as customers.
- *Feedback* by the exploiters is incorporated into the next revision of design and plan.

Fig. 8.9 summarizes the documentation needs and other deliverable of each phase

Fig. 8.10 shows the different high-level phases of a Portal project from the first idea to delivery to the customer. Most of the agile methods focus on the construction phase, as this is where the project usually spends most of the money and flexibility becomes most important. Naturally, the inception and construction phases are equally important. It is usually a smaller team that is working during these first two phases, but their decisions have great influence as they set the stage for the project, by outlining goals, scope, timeline, and team setup.

8.6.4 Integrating Test and Development

As we've pointed out previously, collaboration between development and test is the key for achieving a usable level of quality throughout the release project.

Different tests need to be considered:

- *Build verification testing* and *basic function verification testing* are done for each product build. This is only affordable if these tests can be automated to a great extent, since they need to be executed on different platforms in parallel every day (ideally, all platforms that the product supports). It is very important that basic failures, for example those in install or configuration tasks, are found before a larger number of testers run into the same issue. Build verification testing is provided as a centralized service by the build team.
- *Functional verification testing* is done within each tiger team and is an integrated part of *any* development activity. Design documents with use case descriptions already consider test scenarios and test planning. The testing includes automated regression testing and needs to happen throughout the iteration. There is a risk of the team back-loading their tests towards the end of the iteration, which makes it difficult to identify quality issues early enough to react in time.
  The quality at the end of an iteration needs to be "customer-ready." The follow-on tests should not find any functional problems in the code delivered by the tiger team, especially not to the extent that this would require one of the extended test teams to upgrade to a new build too often to finish their test cases.
- *Extended functional verification testing* takes place in parallel to the ongoing iterations. This testing is done by a dedicated test team independently from the tiger teams. Main focus is extended platform coverage and additional regression testing. The timing of this extended testing is illustrated in Fig. 8.11.

Phase	Vision	Outline	Multiple Iterations (4 weeks each)		Closure
			Detail	Implement and Test	
Activity	Shape the priorities of the Portal organization	Establish teams, which will gather and prioritize use cases	Teams will shape their deliverables in order of priority: detail, implement and test use cases		Regression testing and defect fixing (**overlap with iterations**)
Key Driver	Leadership Team	Nominated Managers/Architects	Nominated Teams	Nominated Teams	Extended Testing
Deliverables	Overal **Goals** and directions (e.g. powerpoint) Nomination of Managers/Architects	**WHAT?** **Charter** prioritized list of all use cases (=overall „Product Backlog" of Line Items), names of stakeholders **HOW?** **High Level Design Document** (i.e. user cases/architecture) Nomination of **Teams**	Updated Charter Updated High Level Design **WHEN?** **Iteration plan** with use cases for this iteration (=current „sprint backlog") **Prototyping Design** (i.e. technical decisions, detailed test scenarios) **Test cases** (i.e. automation)	Updated Iteration plan („sprint backlog list") **Implementation** use cases **Testing** Automated Unit Tests, Functional Testing Performance and System Testing **End-user documentation** (i.e. customer view on use cases) **Iteration Exit checklist** („quality certification" for each delivered use case) **Integration Fest** (show what is „real")	**Extended Testing:** Automated Unit Tests, Functional Testing Performance and System Testing

Fig. 8.9 Documentation and deliverables

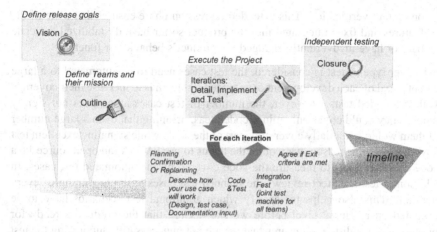

Fig. 8.10 Timing within each iteration

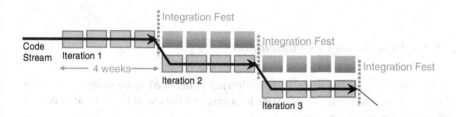

Continues integration and continuous bringup with daily stable drivers

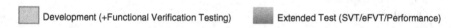

Development (+Functional Verification Testing)   Extended Test (SVT/eFVT/Performance)

Fig. 8.11 Development and test: integration fest is a hand-over point

- *System verification testing* is done in parallel to the ongoing iterations by an independent test team. This testing covers long-running load test scenarios as well as testing in complex environments such as clustered setups, or integration testing with other products under load. The aim is to identify potential memory leaks or instabilities early.
- *Performance testing* happens throughout the release project. The aim is to identify performance regressions and to help the development teams analyze any performance bottlenecks in the implementation. Having stable drivers with no functional problems at the end of every iteration is very important to be able to do performance tests as well as profiling analysis early in the cycle. If this is not done, performance fixes would be rushed in at the end, shortly before release, something that could potentially destabilize the product again.
- A *"release closure"* phase is appended once the last iteration is completed. In this phase, all test teams conduct further regression testing with a special focus

on system verification. This extended regression phase ensures that none of the features and fixes integrated into the product so far have destabilized the code base or have inadvertently changed the product's behavior or function

Whatever type of test is considered, the test cases need to be automated to a large extent. Goal of each development or extended test team is to increase their coverage with automated tests. However, the number of test cases alone does not signify much. They could cover only minor code areas, meaning that even a large number of them would potentially cover only part of the code, while signaling excellent test coverage. For this reason, we require the teams to use EMMA (an open source Java code coverage tool) to measure the code covered by the automated test cases. In addition, a test architect reviews the planned test cases at the beginning of every iteration. This also helps to identify which additional test scenarios have to be executed on a release level before we can declare that the product is ready for primetime. In addition, it ensures that we are not unnecessarily duplicating the test efforts by executing the exact same test cases again on the release level.

8.6.5   Designs and Documentation

At the beginning of the project, the following items need to be designed, written down, and communicated to everyone to ensure that the teams know what to do and where the journey is going:

- Project goals
- High-level requirements
- Overall concept
- High-level product design
- General nonfunctional requirements
- Project user stories.

There are a number of things that can evolve over time or be designed at the time they are first needed, such as:

- Detailed user stories
- Individual tasks
- User interface design
- Supported languages.

The *project goals* are set with the project sponsor and are communicated to the team as part of a kick-off meeting. The kick-off meeting outlines the agile development approach used during the project.

During the release planning meeting, the team and the project sponsor walk through the *high-level requirements*, document them and jointly create the *project user stories*. It is important that everyone has the same understanding of what should be achieved and what the overall project user stories are.

At the same time, the majority of the general nonfunctional requirements are set, such as which Java version, application server, operating systems, and database systems to support. This clarifies the general environment in which teams need to operate.

Based on the *project user stories* and the *high-level requirements*, the team creates the *high-level product design*. This defines the design principles and basic architecture of the project as well as the code standards or guidelines.

In the case of a larger product that requires multiple teams, the agreement on the number of teams and their work split are based on the project user stories, together with the high-level product design.

The project user stories can be detailed at the beginning of the iteration for which they are targeted. At that time the user stories can be broken down into individual tasks.

User interface design, themes, and skins can be developed as the project goes on and as those designs are needed. It is still important that there is a consistent user experience. Therefore we would recommend giving the complete user experience/ user interface design for the overall product to one team of designers as one centralized task.

In projects that will potentially be translated into other languages, all code needs to be enabled for translation from the start, since doing this later on is very cumbersome. Into which languages the product will be translated in the end is something that can be decided later and could usually be done independently of any development effort. This of course assumes that the enablement for other languages is without bugs, which is unfortunately often not the case.

Designs need to be reviewed with all stakeholders and especially by those teams that are in any way impacted by that design. The challenge is usually to get the right set of people together. The owner of the design will put some thought into identifying these reviewers and get them to participate in the review session and sign off the design. The designs also take care of any considerations regarding installation, migration, configuration, globalization, and requirements on user experience, as well as point out any changes in behavior or function compared to previous versions of the product.

8.6.6 Managing Tiger Teams

We have seen how Portal has extended its component oriented organization into a two dimensional model of tiger and component teams. We kept our hierarchical management structure in place, which is focused on functional components and usually consisted of several first line units per geographical location. Developers and testers are grouped based on their component affinity (Fig. 8.12).

In addition to that, we assigned those component team members to tiger teams in order to focus on new product capabilities. For each tiger team we appointed a project lead, who moderates the scrum meetings and prepares the weekly status

Fig. 8.12 The original
structure we used before
switching to agile feature-
oriented teams. Each team
owned their part of the
product code and delivered
their particular piece required
for implementing a feature in
the product, with none of
them really being responsible
for the overall integration of
the feature

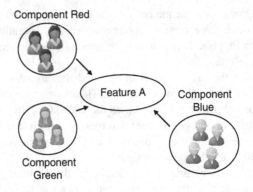

Fig. 8.13 In this new team
structure, the team focusing
on a feature is staffed with
experts from the component
teams

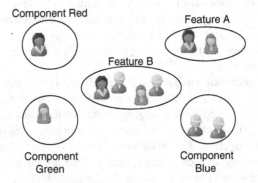

reports. We also assigned an experienced software architect to each tiger team, who
ensures technical soundness of the feature being developed. It turned out to be quite
helpful to also add a first line manager as a sponsor or godfather for each team.

One major challenge for the team is to create a common team spirit across
locations and different hierarchical reporting structures (Fig. 8.13). Another challenge is to engage the stakeholders and ensure that their needs are reflected in the
product backlog

Beside the component and tiger teams, there is still an overarching release
management team, which comprises the overall technical and management leadership. In our case we had release manager, release architect, chief programmer, lead
architect, as well as representatives from product management and executives
(Fig. 8.14).

The tiger teams have full responsibility for the development of their feature and
the freedom to organize their team in the most efficient way. But they have to fit
their work into the overall release framework, which, for example, defines the same
iteration length as well as iteration start and end dates for all the teams.

Even with distributed teams, it is good to have daily 15 min scrum meetings to
exchange what everyone in the team is working on and which new issues have

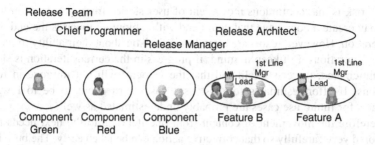

Fig. 8.14 Project structure with components, feature teams, and release team, which is managed in a matrix structure orthogonal to the hierarchical organizational structure that is already in place in each location

Iteration	Start	End	Dev	Test		Status
				%A	%C	
Iteration 0	7-Jan-08	15-Feb-08				
Iteration 1	18-Feb-08	28-Mar-08				
Iteration 2	31-Mar-08	9-May-08				
Iteration 3	12-May-08	20-Jun-08				
Iteration 4	23-Jun-08	1-Aug-08				
Iteration 5	4-Aug-08	12-Sep-08				
Iteration 6	15-Sep-09	10-Oct-08				

Test Automation Current Iteration	Number of **Planned** automation test cases	Number of **Delivered** automation test cases
Rational Function Tester	TBD	TBD
Rational Perf Tester	TBD	TBD
Total Code Coverage	TBD	TBD

Team name	Overall Status
DATE:	**Green/Yellow/Red**
Highlights	
•	
Key Issues	
Risks	
•	
Required Actions by Senior Management	

Fig. 8.15 The figure above shows the overall status template we are using for the weekly release status meetings

appeared. To aggregate this team specific information to the overall project status, we schedule a release status meeting ("Scrum of Scrum") once a week. Fig. 8.15 shows a example status chart used for this purpose.

The status of the current iteration should show the status against all user stories planned for the current iteration. If a team realizes that it cannot complete all items originally planned for the current iteration, it needs to highlight the user stories at risk and mark their status as being yellow or red. Those user stories that cannot be fully completed and tested in the current iteration will be returned to the product backlog. They may be picked up again for the next iteration.

One risk is that continuous movement of user stories from one iteration to the next may create a bow wave that could end with a blow-up towards the end of the last iteration. However, teams are often too optimistic about their ability to recover in future iterations. But keep in mind: if progress in the current iteration is slower than anticipated, there is likelihood that the velocity will not improve in future iterations. If effort estimates for past uses cases turned out to be to low, the estimates for future use cases are probably underestimated as well.

Therefore the movement of content from one iteration to the next needs to be monitored very carefully so that corrective action can be taken early. The burndown chart for the remaining use cases within the overall product backlog will indicate, if the remaining committed items can still be addressed in the given timeframe of the release project.

8.7   The Benefits and Pain Points

In summary, agile practices proved to increase the flexibility of the development project and especially the efficiency within the development and test organization.

Nevertheless the move from the waterfall model towards tiger teams and iterations has been a challenging journey. The WebSphere Portal development team has mixed and matched suitable ideas from the palette of known agile techniques and concepts. Part of this journey has always been to reflect the applied techniques and adapt Portal's agile approach where it becomes necessary.

In the following, we've summarized our experience with WebSphere Portal into a set of guidelines that will hopefully provide you with a short-cut to a successful agile project:

- *Get the commitment of the entire organization:*
  Agile software development requires the will to change be flexible of all people involved. Agile needs *commitment by the entire organization* in order to be successful. The *leadership team* needs to trust and empower the teams. They need to give guidance by communicating a well-defined overall vision, high-level requirements and a rough timeline in which deliverables are expected. And they need to encourage a common team spirit within each tiger team. Without this support, Portal would never have managed to move away from the waterfall model.
- *Establish tiger teams and make them accountable for their results:*
  Each *tiger team* is accountable for its results. Everyone needs to understand what other team members do. Teams need to consider risks, take over responsibility, and be prepared to come up with mitigation plans quickly. Tiger teams span organizational structures and locations. They drive customer-oriented, end-to-end use cases with fewer gaps in between. They think out of the box rather than being limited by a component-centric point of view. This has turned out to be the key strength of Portal's agile approach. This kind of teaming brought together individuals from different locations and different disciplines.

- Apply a *two-level planning:*
  Do not waste time with planning exercises on too granular a level. The project status 6 months from now is not predictable. Be patient: the first iterations will give you a well-founded assessment of progress, status, quality, and remaining capacity of the teams. This kind of information will be a much more reliable baseline for ongoing fine-grained planning.
- *Plan for change:*
  Commit only to very few and most important high-level use cases initially. The team can add more once the product is on its way and there is less planning ambiguity. This core aspect of agile thinking has been extremely controversial and difficult to implement, as it is a clear contradiction to the teams' and management's desire to understand well in advance the work and issues lying ahead. A release manager is accountable for the success of a project, after all, and will therefore want to know the details far ahead of time to be sure that the deliverables are on track.
- Find the right *balance between flexibility and planning:*
  Early commitments will narrow down your future options. On the other hand, you need to commit a certain level of functionality to your stakeholders to get their agreement to even start the project. Keep in mind that flexibility is good, but is not for free! It is crucial to define the major focus areas for the investments within the release project. A rough timeline is a must. It needs to be done early, whereas finalizing the exact feature scope or the shaping the exact out-of-the-box functionality can be done rather late. Achieving the right level of commitments at the right time requires experience. It took us two releases to find a reasonable balance.
- Do the right *amount of documentation:*
  Planning and documentation are needed to document agreements and commitments, to describe the architecture, and to specify interfaces. Do not use "Agility" as an excuse for lack of planning and design preparation. We started agility by banishing virtually everything written, but quickly reverted to a stricter approval process of formal documents. In the end, we produced a reasonable set of documentation without any underlying formal approval process.
- *Break down big pieces of work into smaller use cases:*
  Make the pieces a size that can be implemented and tested within a single iteration. Finding the right granularity for use cases has turned out to be difficult to achieve. The work items were far too often bulky and hard to manage.
- *Focus on working code and quality from the very first iteration:*
  You will see immediate benefits when continuously integrating small changes and when putting the developers in charge of quality. Working code lets you to run comprehensive stress, load and performance tests during the iterations, instead of delaying them to the very end of the overall release, at which point it would be too late to fix significant issues anyway. Working code also helps to understand the actual progress. Frequent and measurable results are an essential piece of information for fine-tuning the planning of the next iteration.

Establishing a mindset which prioritizes the fixing of problems higher than adding new functionality required a significant amount of evangelizing on all levels of the project. But an early focus of reducing the amount of unresolved bugs, helped to avoid the rush at the end of the release.

- *Automate testing:*
This avoids regression problems when the next set of use cases is implemented in a subsequent iteration. In our project, coverage of automated test cases has been growing, but the adoption rate has been quite slow. New features still tend to have priority over a more extensive test coverage.

- *Have an "Integration Fest":*
The *"integration fest"* demonstrations proved that the teams accomplished working code by continuous integration throughout the project. These demos proved extremely valuable as they produced solid baselines for beta deliverables and extensive testing.

- *Involve Customers early:*
The product backlog needs to reflect what the customer really wants. Therefore, in most agile approaches, this backlog is not owned by the developers, but directly by marketing or product management. However, in our case we asked the tiger teams to assume this ownership themselves and solicit the stakeholders for input on prioritization.

Chapter 9
Summary and Wrap-Up

9.1 The Essence of Agile

To become agile, it is not sufficient to just install a handful of new tools, apply some new methods, and rename your milestones to "iteration exit." Agile will challenge the way your organization is set up, and it will affect the daily work of each individual. Thinking horizontally across boundaries of teams, geographies, and organizations is a change of culture and mindset. Part of embracing your highly dynamic environment is to accept that you will learn as you go by prototyping and evolving your deliverables over time. Change is always part of the game. Ask for feedback and adopt! You will continuously integrate your work, while continuously stabilizing and evaluating whatever you are doing – regardless of whether you are planning, designing, or implementing. Staying focused on customer value and proceeding in a sustainable pace will drive efficiency.

Yes, agile development is a lot about philosophy, and it needs the commitment of the entire organization. Adopting agile will take a lot of hard work and considerable time until you can reap its benefits.

Over the past 200 pages, we have covered many aspects of agility. We have underlined them with considerations and experiences. But what is the net of all this? If you had to summarize what agile software development is all about in a few bullets, what would they be?

- *Keep it simple!*
  Develop incrementally in small, time-boxed iterations of a maximum length of 4 weeks:
  This forces everyone to break complex problems into really small chunks of work that can be understood and managed. You will gain simplicity and will make fact-based decisions rather than wild guesses. Do not overcommit! In such short iterations there is no opportunity for feature creep, as you are forced to stay focused on your fundamental objectives. The pace of a project will be a steady rhythm of continuous planning, designing, implementing, testing, and

T. Stober and U. Hansmann, *Agile Software Development*,
DOI 10.1007/978-3-540-70832-2_9, © Springer-Verlag Berlin Heidelberg 2010

delivering. The rough project outline will guide the teams into the future, while the precise elaboration of details will focus on the present only.

- *Team with the customer!*
  Get stakeholders involved:
  This forces you to ensure that you meet expectations and deliver customer value. Continuous feedback will provide an early validation of whether you are doing the right thing and how it could be further improved to meet the customer's expectations.
- *Show and tell!*
  Working code is the bar to measure progress:
  This forces you to focus on quality. Delivering working and immediately tested code will provide an early verification of whether a use case is really done. In this context, it is crucial to heavily use test automation. Having the entire code base functioning at all times implies that a project is well under control and is able to focus on value rather than struggling with permanent chaos. If you continuously integrate and deliver tested code changes, you avoid letting issues accumulate until they go out of control.
- *Empower the teams!*
  Establish autonomous "fractal" teams as the source of value:
  This forces you to lower the center of gravity and let those who are the closest to the object of discussion make faster decisions. Leaders will coach teams like a Jazz session rather than directing them like a classical orchestra.

Recall the five key principles of fractals, when you unleash the full potential of teams:

- *Self-similarity*: All teams are alike and operate within the same corporate culture and vision.
- *Goal-orientation*: A team derives its scope of action from an agreed set of high-level goals and strategy. The team has the necessary authority to pursue its goals.
- *Self-organization*: Teams organize their work in a way which suits them best to accomplish their mission.
- *Self-improvement*: A team continuously improves and adapts its products and processes based on reflections and feedback.
- *Vitality*: Extensive interaction, effective collaboration, flexible planning, minimalist processes, and a lean organization without a rigid confinement boost the team's efficiency and help to find a creative solution when the unforeseeable happens.

Anything beyond these fundamental points can be your individual mix of practices that you believe will improve you project execution. Remember, there is no general rule to find the right balance between flexibility and structure. Getting to the right level of detail, commitment, documentation, autonomy, and simplicity is an ongoing process that will evolve in parallel with your product development.

9.2 Building an Agile Community

Agility is strongly influenced and shaped by practitioners. It can therefore be extremely helpful for developers to organize themselves in communities.

"Agile@IBM" is a very active community within IBM that offers a networking platform to exchange experiences and ideas. The group organizes internal conferences, workshops, and education materials. Approximately 900 members from all IBM divisions and geographies are actively influencing IBM's vision and implementation of software development.

The Agile Alliance is an open community that evangelizes the values of the manifesto for agile software development. That group offers valuable materials and events and invites agile practitioners to join.

9.3 Comparing once again

Throughout this book, we have kept bashing on the plan-driven waterfall approach, which we tend to position as heading straight ahead without looking to the right or left. It is high time to apologize for having carried this to extremes. We are in fact

Table 9.1 Comparing teaming aspects

Aspect	Traditional Development Paradigm	Thinking Agile
Management Style	Hierarchy and control	Management by objectives
	Centralize decisions	Decentralize and delegate decision
	Direct	Lead and coach
Organization	Fixed structure	Adaptive, emerging structures
	Function-oriented structure	Teams have end-to-focus
Working Style	Mistakes are failure	Mistakes show ways for improvement
	Control	Trust
	Guide	Collaborate
	Culture of sign-off and approval	Shared learning
	Weekly status meetings	Daily stand-up meetings
Responsibility	Component-specific view	Holistic approach
		Team spirit
	Team members deliver	Team members are stakeholders
	Teams fulfill assigned tasks	Teams assume responsibility
Collaboration	Limited communication within team	Unlimited communication within team
	Authoritarian environment	Collaborative environment
Customer	Contract negotiation	Customer collaboration
Tools and Processes	Overly complex tools	Easy-to-use tools
	Processes and tools rule the project	Individual and interaction go first
	Rigidity of tools and processes	Flexibility of tools and processes
	Complex and comprehensive processes	Minimalist process and practices

using traditional project management as a metaphor to explain the characteristics of agile software development. Agile and waterfall are probably at opposite ends of the broad variety of project management doctrines. The comparison of the two helps to point out and illustrate the core aspects – even if there never has been or never will be a single project that embraces either extreme to the full extent.

We have taken you through the numerous differences between agile thinking and traditional project management. In the tables below, we compare and summarize the key differences that this book covers (Tables 9.1–9.4). Fig. 9.1 summarizes the key phases of an agile software development process.

It is a pleasure for us to have you as one of the readers of our book, and we hope it has provided you with additional value that makes it easier for you to make the right decisions in your current or next project using agile software development techniques.

We would be more than happy to hear from you and your experience with agile software development. You can reach us via tstober@gmx.de or uwe@hansmanns.net.

Table 9.2 Comparing planning and designing aspects

Aspect	Traditional Development Paradigm	Thinking Agile
Detail	Complex and comprehensive planning	Evolving plan
	Plan for entire project	Two-level planning:
	Detailed long term schedule	Rough outline for project (prioritized high level goals, team budgets, overall schedule)
	Detailed long term schedule	Fine-grained plan for current iteration
Estimating	Estimated by small team of architects	Collaborative estimating; Planning poker
	Person days	Story points or ideal days
Timing	Detailed planning early	Detailed planning late
	Project phases and milestones	Short iterations
	Protect the scope and requirements	Protect the date (time-boxed)
	Fixed scope	Prioritization of scope
Plan owner	Plan created by project management	Plan is jointly established by overall team
	Top-down planning	Bottom-up planning by self-managed sub-units
Plan changes	Decided set of requirements from the start	Vision and prioritized backlog
	Rigid definition	Outline
	Plan Change Process	Reprioritization
Communication	Limited/restricted access to the plan	Organized/unlimited access to the plan
Design	Specification at start	Just-in-time design elaboration
	Rigid definition	Prototyping
	Architecture is planned	Architecture emerges

Table 9.3 Comparing implementation aspects

Aspect	Traditional Development Paradigm	Thinking Agile
Responsibility	Function-oriented work split	End-to-end use case orientation
	Individual or component specific code ownership	Collective code ownership
Coding	Integrate late/milestones	Continuous integration of small incremental iterations
	Independent component deliverables	Single code stream
	Handover to test teams	Immediate testing
	Function first	Quality first
	Test automation is optional	Development of automated test suite
Measure of Success	Conformance with plan	Working code
	Demonstrate at the end	Always running code
	Complete development work first	Never break the build
	Compliance with specification	Customer satisfaction

Table 9.4 Comparing testing aspects

Aspect	Traditional Development Paradigm	Thinking Agile
Ownership	Independent of development	Part of development activity
Timing	Test late	Test immediately
Responsibility	Test team is responsible for test	Everyone is responsible for test

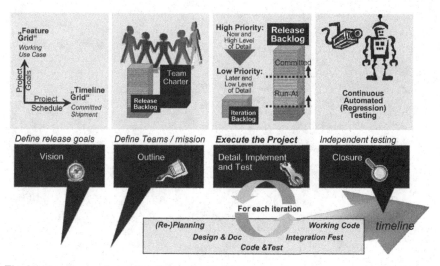

Fig. 9.1 Key phases of an agile software development process

Index